Encyclopedia of
Sustainability

Encyclopedia of
Sustainability

Equity and Fairness
VOLUME III

Robin Morris Collin

Robert William Collin

GREENWOOD PRESS
An Imprint of ABC-CLIO, LLC

A B C ☰ C L I O

Santa Barbara, California • Denver, Colorado • Oxford, England

Copyright 2010 by Robin Morris Collin and Robert William Collin

All rights reserved. No part of this publication may be reproduced, stored in a retrieval system, or transmitted, in any form or by any means, electronic, mechanical, photocopying, recording, or otherwise, except for the inclusion of brief quotations in a review, without prior permission in writing from the publisher.

Library of Congress Cataloging-in-Publication Data

Encyclopedia of sustainability / Robin Morris Collin, Robert William Collin.
    3 v. cm.
  Includes bibliographical references and index.
  Contents: vol. 1. Environment and ecology —
  ISBN 978-0-313-35263-8 (vol. 1 print : alk. paper) — ISBN 978-0-313-35264-5
(vol. 1 e-book) — ISBN 978-0-313-35265-2 (vol. 2 print : alk. paper) —
ISBN 978-0-313-35266-9 (vol. 2 e-book) — ISBN 978-0-313-35267-6
(vol. 3 print : alk. paper) — ISBN 978-0-313-35268-3 (vol. 3 e-book) —
ISBN 978-0-313-35261-4 (set - print : alk. paper) — ISBN 978-0-313-35262-1
(set - e-book)
  1. Environmental sciences—Encyclopedias. 2. Sustainability—Encyclopedias. 3. Sustainable
development—Encyclopedias. I. Collin, Robin Morris. II. Collin, Robert W., 1957–
  GE10.E528 2010
  333.7203—dc22   2009037029

14 13 12 11 10   1 2 3 4 5

This book is also available on the World Wide Web as an eBook.
Visit www.abc-clio.com for details.

ABC-CLIO, LLC
130 Cremona Drive, P.O. Box 1911
Santa Barbara, California 93116-1911

This book is printed on acid-free paper ∞

Manufactured in the United States of America

CONTENTS

GUIDE TO RELATED TOPICS

PREFACE

References to sustainability are everywhere, from advertising to space travel. Words associated with sustainability are fast becoming ubiquitous. This reference text is designed to help understand the many meanings of sustainability.

Concepts of sustainability have been developed in multiple disciplines including the sciences, international agreements, development law and policy, and humanities. The essential concepts about sustainability can be described in terms of three broad domains:

- Environment and ecology

- Business and economics

- Equity and fairness

The relationship between these domains is described in somewhat different ways. Some describe their relationship as a three-legged stool or three intersecting circles. Each circle or leg of the stool represents one domain, environment, economics, and equity. Each circle is of equal size; each leg bears equal weight.

Others describe the fundamental relationships as three nested baskets. The environment is the largest, most comprehensive basket. Within it, all human activity is located including human economic enterprises and human communities. Economic enterprise is represented by a basket nested within our environment and its webs of life. Human individuals

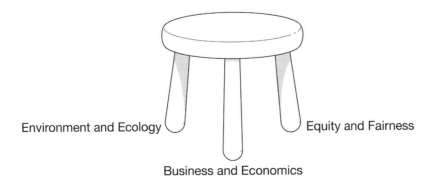

Environment and Ecology   Equity and Fairness

Business and Economics

Figure 3.1 • Sustainability as a three-legged stool: Environment, economics, and equity. Illustrator: Jeff Dixon.

and their communities rest within both these two baskets relying on each for their livelihood and support. This encyclopedia set will provide the reader with the necessary infrastructure to navigate the complex roads and byways of the contemporary discourse on sustainability. That infrastructure is based on the axes of environment and ecology, economics and business, and equity or fairness. Sustainability weights these three areas equally and joins them in every discourse. This triangulation of the so-called three "E"s—environment, economics, and equity—distinguishes sustainability as a philosophy different from that of conservationism or environmentalism. This encyclopedia devotes one volume to each of the three "E"s.

In each volume, there is the same basic organization of chapters: a comprehensive introduction, definitions in contexts that are pertinent to that particular volume, the contemporary public policy contexts arranged from the global to national to local levels, current controversies, and future trends.

Each volume begins with a comprehensive overview of what the term *sustainability* means in each domain. This comprehensive overview introduces the major concepts of sustainability as used in each unique

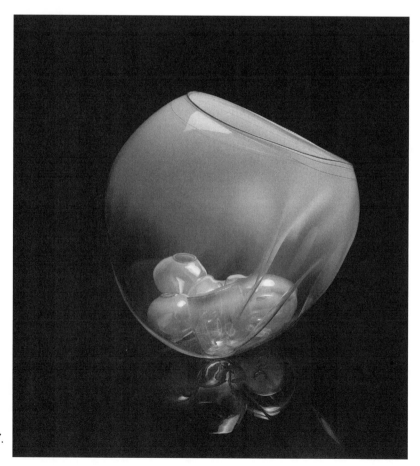

FIGURE 3.2 • Sustainability as three nested baskets: Environment, economics, and equity. Copyright © Dale Chihuly, 1992. Sky Blue Basket Set with Cobalt Lip Wraps, 1992 17″ × 15″ × 16″. Photo by Terry Rishel.

volume: environment and ecology, economics and business, and equity or fairness. This introductory chapter provides a concentrated account of the "big picture" in each unique arena of sustainability. The interconnectedness of basic ideas makes the study of sustainability challenging to the novice and complex to anyone. The overview section of each volume presents basic concepts and relationships in the primary context of the volume. The overview provides concise insight into terms of art, the dynamics that have shaped the concept within that context, and the changes and challenges that sustainability presents in that particular context. These are the elements of complex interactions and the background on which human choices and policies act and interact with natural systems.

Following the overview, each volume contains a chapter on definitions and contexts that give in-depth descriptions of key terms set into contexts relevant to that particular area. This chapter is a primer on the basic terms and definitions that are foundational to each volume. Without such a primer, terms of art can become a private language of expertise, making knowledge and information inaccessible to the general public reader, even one with considerable education. This section identifies key terms of art and defines them in accessible language. Definitions are arranged contextually, and alphabetically. This organization is tailored to assisting the reader with quick and ready access to the language and contexts of the sustainability discourse. This chapter gives a reader ready access to the specialized terms of art and foundational definitions of the discourse in each area. The dynamic interaction of these terms is illustrated in images and examples throughout this volume.

Next in each volume Chapter 3 describes the role of government and the United Nations in achieving sustainability. We include a description of United Nations programs while recognizing that the UN is not a government but an association of sovereign governments. The UN has exercised global leadership in guiding world governments toward sustainability. We also describe the work of national, regional, and local governments. Local governments have a uniquely important role to play in implementing sustainability because they are most closely connected to place and community. Even in countries whose national governments have chosen to abdicate their role in achieving sustainability, local governments have acted independently as laboratories and as activist organizations to achieve important behavioral changes. Many United Nations programs and policies are aimed at this local and regional level of government as well as through nongovernmental organizations (NGOs).

In the government section of each book, public policy is arranged in a global to local progression and, within that structure, it is presented in its chronological order. We have organized each chapter on government involvement to trace the historical developments as they have occurred at different levels: global developments through the United Nations organization, U.S. national developments where they have contributed, and the state, local and regional efforts within the United States. Sustainability public policy is rapidly unfolding in these different venues at uneven rates of change. For example, U.S. municipalities are rapidly adopting the Precautionary Principle in land use ordinances designed around sustainability.

In Chapter 4 of each volume, controversies related to sustainability are explored in greater depth. Every environmental and developmental conflict reflects competing and sometimes conflicting interests of many interested parties. Too often, these are portrayed in the popular media as simple conflicts between two parties. The truth about controversies, however, requires an appreciation of the competing and conflicting interests of multiple stakeholders. In the end, all solutions will be local, as are all environmental and ecological controversies. Solutions are not the focus of this section of the volumes here. Instead, we aim to describe the nature of the interests involved in all aspects. Controversies are once again arranged alphabetically. Each volume lists and describes current controversies in each volume. These controversies were selected based upon their political salience, and likely impact upon future generations. Controversies around sustainability provide a rich area for classroom discourse.

Finally, each volume concludes with a section devoted to emerging trends arranged alphabetically. This chapter takes a considered look at trends and data to follow for future developments. These trends were selected partially based upon the availability of existing data collections, and the commitment of governmental organizations and nongovernmental organizations to collect and monitor data. New sources of data and better resources will continue to develop but these fundamental trends should lead the interested reader to the sources that we have now. In this section, we describe what the future of sustainability in each area may hold based on the information available now. At this moment, some changes seem inevitable, whereas others may be subject to human management. The future of our relationships built on a sustainable model of dynamic and radical inclusion is the subject of imagination and possibilities.

The concept of sustainability is divided into three co-equal components of environment, economy and equity. We did so to arrange the large amount of information in a manner most comprehensible to the reader. These components of sustainability are dynamic as well as interrelated. We developed our framework of overview, definitions, government involvement, controversies, and future trends per volume to help the reader understand sustainability in the context of each of the co-equal components. Within each framework, we have further introduced concepts of scale, such as going from global to local levels of government intervention. When appropriate, we have introduced chronologies that underscore the development of sustainability. Each volume is unique, and capable of standing alone, but with a similar framework. Each volume is cross-referenced, with portal Web sites. Our goal is to provide a comprehensive framework that the reader can easily navigate within and between volumes.

The ultimate challenge of human sustainability is how human enterprise and communities that can establish themselves without undermining the fundamental health of the natural systems on which all beings on Earth rely. The legacy of previous centuries, with their development of human enterprises and communities of great scale reliant on diminishing ecological resources, is the presence of wealth amidst poverty and environmental degradation that challenges our ability to survive. Sustainability as a doctrine challenges us to provide for our needs while allowing future generations the same opportunities for prosperity and a full experience of life. The idea that we can do that for future generations while ignoring the growing inequities of contemporary life is an equal challenge to sustainability. The central challenge of sustainability is how human enterprises and communities can function within ecosystems supporting our environment. The great human progress and development of the contemporary era have been accompanied by a growing gap between rich and poor people in the context of widespread environmental deterioration, and increasing poverty. Earth's ecosystems provide enormous benefits for humankind. Some of those benefits are from renewable resources, and some are not renewable. Humankind has exceeded the limits of some renewable resources and is approaching the limits of nonrenewable ones.

For those who are interested in the idea of sustainability and wish to explore it further, there is an overwhelming volume of material to read devoted to specific contexts and applications. Often, this material is difficult to penetrate for a novice because it is so heavily reliant on specialized language and specialized constructs unique to a particular discipline. Theses volumes provide a gateway that allows access to the full variety of the field and facilitates independent investigation in a multidisciplinary field. Sustainability will require new ways of thinking about the environment and a basic shift in public policy at all levels of government. This reference is dedicated to the task of facilitating human imagination and thought in that direction. Imagination is a uniquely human faculty praised in physics and metaphysics alike. Albert Einstein said that imagination was more important than information. Buddhism insists that thought and intention are as important as the acts to which they may give birth. Creativity may be inspired by the interplay between the major perspectives offered in each volume. Pragmatic implementation is illustrated in stories, biographies, and illustrations throughout these volumes. These are designed to help the reader understand key concepts, and thereby provide a springboard for the next generation of human imagination and sustainability.

References

Anderson, William. 2001. *Economics, Equity, Environment.* Washington, DC: Environmental Law Institute.

Capra, Fritjof. 1996. *The Web of Life: A New Scientific Understanding of Living Systems.* New York: Anchor Books.

Collin, Robert William. 2007. *Battleground: Environment.* Westport, CT: Greenwood Press.

Collin, Robin Morris, and Robert William. "Where Did All the Blue Skies Go? Sustainability and Equity: The New Paradigm." *Journal of Environmental Law and Litigation* 9 (1994):399–460.

Dubash, Novraz K., and Daniel Bouille. 2002. *Power Politics: Equity and Environment in Electricity Reform.* Washington, DC: World Resources Institute.

Johnson, Steven M. 2004. *Economics, Equity and the Environment.* Washington, DC: Environmental Law Institute.

Paehlke, Robert. 2008. *Democracy's Dilemma: Environment, Social Equity, and the Global Economy.* Cambridge, MA: MIT Press.

ACKNOWLEDGMENTS

We would like to express our gratitude for all those who helped us with this encyclopedia. Willamette University's President Lee Pelton and the Center for Sustainable Communities provided foundational support. We are also grateful to David Paige at ABC–CLIO Press, for his support, patience, and timely assistance. We are grateful to the students who worked as research assistants for us in this process: Sikina Hasham, Lacey Lucas, and Sarah Hunt Vasche. Candace Bolen provided invaluable office support for us for which we are grateful. We would also like to thank all our students over the years. More than a decade of teaching the first sustainability course in a U.S. law school, several sustainability courses in university environmental programs in the United States and abroad, and environmental justice issues to communities and government agencies has exposed us to a wonderful cohort of earnest, thoughtful, and hopeful students. These students are from Auckland University, New Zealand; Cambridge University, UK; Urban and Regional Planning, Jackson State University; Department of Urban and Environmental Planning, University of Virginia; Environmental Studies and Law school at the University of Oregon; Willamette, Tulane, and Lewis and Clarke Law Schools, Hunter College, and Cleveland State University School of Social Work. Many communities and indigenous peoples have also shared their visions of sustainability with us. We are very grateful for their assistance. They include the Choctaw Band, the Spokane Tribe, and the Indigenous Environmental Network. Many community groups and their leaders have shared their hopes for justice and sustainability. They include the Environmental Justice Advisory group, Center for Community Environmental Justice, and many urban community leaders. The federal and state environmental agencies also have their share of students we are grateful for. The U.S. Environmental Protection Agency, the Oregon Department of Environmental Quality, the Washington Department of Ecology, and the Missouri Department of Environmental Quality all contributed students of sustainability. All of our students provide us with a window to the future, a future they want to be sustainable.

INTRODUCTION

Sustainability is an important concept now developing into public concern and policy all over the world. It challenges us in many ways. It challenges us to account for all environmental impacts, past, present, and future. It also challenges us to examine closely how we treat each other, as well as how we treat the environment. It challenges our fundamental values as to what is fair. As sustainability becomes implemented in policy and behavior, questions about fairness arise. What products get to be labeled as "sustainable?" What public participation processes are "fair" when planning for sustainable land use planning? Should the rich pay more for sustainability because they can and the poor cannot? Ultimately, sustainability is an unfolding dynamic of human processes rooted in an ecologically sensitive framework. Questions are framed in terms of achieving sustainability. . This challenges our patience and our understanding. This encyclopedia is written to increase understanding of the concept of sustainability in all its unfolding directions and paths.

The key principles of sustainability as enunciated by Agenda 21 are:

- Integrated decision making

- Polluter pays principle

- Sustainable consumption and population levels

- Precautionary principle

- Intergenerational equity

- Public participation

- Common but differentiated responsibilities

Each one of these principles of sustainability relies on basic principles of fairness in terms of public participation and intergenerational equity. Fairness is one of the most uncompromising foundations of sustainability, giving sustainability proponents and policies impetus to challenge traditions, to change ways of thinking about environmental issues (paradigm change), and to engage in adversarial and controversial issues. Resistance to sustainability will come from some claims of perceived "unfairness." People who own private property expect to use it as they please, even if it is not sustainable. Will the highly held value of private property clash with the new wave of sustainability policies?

Most cultures were aware that their long-term survival was reliant on natural systems. Only recently have all cultures been made aware that there are limitations on natural systems with the advent of global warming and climate change. More and more people across various stakeholder categories of industry, government, community, environmental, religious, other nongovernmental organizations and labor are able to observe human impacts on the environment. Knowledge of the effects of these impacts on humans and on the ecosystem is slowly emerging in many different forums. Science is often necessary to prove causality and can often prove that a type of behavior either caused or did not cause an event. (the equivalent of saying you cannot confirm or deny causality). Other community observations around impacts on humans do not reach "causality," and it is here that many legal and policy controversies develop. Is it "fair" to make someone change behavior when that behavior may not be causal in environmental degradation or increasing threats to public health? Is it fair to have a community and environment exposed to pollutants when they do not create them?

In this volume, we focus on the equity and fairness issues for sustainability. Issues of environment and ecology and business and economics are covered in the other two volumes. Concepts of what constitutes an "environment" span many cultures, and many cultures suffer from environmental ethnocentrism. Knowledge about the environment is not evenly distributed, and some regions may discover today what others have known for a long time. Many nations consider the environment as everything around them, including themselves. Other nations, such as the United States, do not consider the built, or urbanized, environment as part of the environment.

The globalization of environmental observation, information, nongovernmental organizations, other civil societies, and technological advancement has helped overcome this dearth and disparity of environmental observations. With these increased observations, scientists and others were able to see some of the interconnectedness between land, air, and water natural systems and to relate climate to landscape. Units of complete life systems were called biomes or ecosystems. The globalization of ecological information has been a major push in the social acceptance of sustainability worldwide. Many nations are already seeking ways to reduce carbon dioxide emissions drastically, sequester carbon dioxide, and to accurately evaluate ecosystem carrying capacity.

As governments and communities struggle to make environmental science the basis and parameter for policies around sustainability, knowledge about human impacts on natural systems continues to grow. Knowledge about how to live sustainably will also grow. Questions about fairness that occur when sustainability is real policy may increase the level of justice in society. That is the purpose of this volume.

CHAPTER 1

Introduction and Overview

What I want to see is an environment where the young people of our country have a real chance to develop the inherent possibilities they have to create a better life for themselves . . . That is what development is about.

Nelson Mandela, 2006, The Ambassador of Conscience
Award, Amnesty International (available online at
www.amnesty.org)

WHAT'S FAIRNESS GOT TO DO WITH IT? HOW EQUITY FITS INTO SUSTAINABILITY

Humans can comfortably live on only 20 percent of the Earth. Most live on land, between sea level and one mile in altitude. Fresh water, stable landmasses, and gentle climates are often the point of development of human habitation. The ecosystems in these areas feel human impacts that are the most noticeable, but they are not the only environmental impacts. By the year 2030, 60 percent of the world's population will be living in urban settlements—cities and towns. The majority of people will be living in less developed countries, and 84 percent of people living in developed countries will be living in cities.

The majority of urban dwellers are people of color. Together with people of color, women, and children will compose the majority of the world's population. As the deterioration of our ecosystems accelerates, these people suffer most. According to the Millennium Ecosystem Assessment, commitment to people in these demographics has been historically lacking. The historical lack of commitment to poor people and people of color is a large problem for global approaches to sustainability. As the global systems of nature become less sustainable, these populations are affected more and are needed more by the rest of the planet. Ecosystem degradation affects us all on several fronts.

Food security is a major result of ecosystem degradation. Rural areas of poor countries are the most dependent on local ecosystems for food, but everyone is dependent on some ecosystem for food. Climate changes greatly increase the risks to traditional agricultural production. Currently the world population is slightly less than 6 billion people, with about 800 million overfed and 800 million underfed. Another 1 billion people are malnourished. With population increases, cumulative impacts on ecosystems from industrial agriculture, and climate changes, the threat to food security is increasing.

Water is the source of life, and fresh water is needed for many ecosystems. According to the Millennium Assessment, more than 1 billion people do not have access to safe, fresh water. Only about 7 percent of the water on Earth is fresh water. Underground water supplies in aquifers usually become contaminated before they dry out, as the chemicals and waste become more concentrated. Water-based deaths from infectious diseases account for about 6 percent of global deaths. This is about 1.8 million deaths per year, many of them small children. Fresh water is closely linked to quality of life and to economic development. Many people have water-borne infectious diseases for part of their lives that affect their predisposition to other diseases, which can kill them.

Air pollution from indoors and outdoors is beginning to erode the ecosystems and the public health in urban systems. In the United States, the average asthma rate is about 7 percent, but in urban areas like Portland, Oregon, it is about 14 percent in communities of color. In other countries indoor heating and cooking using biofuels burden the breathing of the population. On the global scale, about 3 percent of disease is attributed to indoor air pollution. Generally, the more concentrated the population, the more concentrated the pollution from this source. Increasing urbanization would therefore increase the risk of airborne diseases. If the source of cooking and heating is biofuels or wood, and this becomes unavailable, then other health issues increase. Without the ability to keep warm or to boil water and thoroughly cook food, other diseases take hold. In many countries, it is the role of women to get the wood and clean water to cook and clean. As deforestation increases, these women must walk farther and longer and lose time and energy to resist disease, nurture children, tend crops, and attend school. Deforestation often leads to decreased water supply because the trees provide shade over water to slow evaporation and because the tree roots retain water and slowly release it.

As human populations increase and become more urbanized, threats from both naturally occurring poisonings and human waste management practices increase. Some naturally occurring water sources may have high levels of dangerous chemicals such as arsenic. In the push for fresh water, some of these sources may be tapped without adequate testing or purification. They may be used directly as drinking water or indirectly as irrigation water. They may also leech into another underground water source if that source becomes empty or low. Human sources of water contamination can result from persistent organic pesticides that can last a long time after their use and stay in underground water systems. PCBs, DDT, and dioxins are common chemicals. Although they may be banned from some industrialized nations, manufacturers and distributors of these chemicals move them into developing nations. Even when they are banned, environmental enforcement is poor. They can cause disruption of endocrine systems in living things. This type of physiological disruption affects animal husbandry, crop productivity, and human health.

Human societies greatly value culture and traditional practices. These vary greatly throughout the world. They include recreation, beauty, tourism, educational, cultural preservation, and a sense of place. With ecosystem erosion, these values are also affected. Some can be translated into economic losses, others into health quality losses. An ecosystem linked to these values and practices is important to sustainable development because changes in human behavior linked to sustainability will be affected by them.

Food, water, energy, and culture are all affected by how the climate changes in response to global warming. Fast changes can be disruptive to ecosystems. Rising ocean levels will cause salinization of fresh water supplies farther inland, and much of the world's urbanizing population is growing near a coast. Direct effects in the near future are most easily known, but indirect effects in the middle or distant future are less known. The lag time between human action and its effect on actual climate change is a major factor in the uncertainty. It is compounded by the lack of knowledge about direct and indirect impacts. Some facts are certain though. Human population growth in urbanized areas and poverty that is ignored will have negative impacts on global environmental systems.

There are two paths to both protect public health caused by eroding ecosystems and to be part of sustainability. Both are expensive and require direct action in the face of uncertainty. This makes them controversial. The first is to prevent or mitigate the actual environmental damages. The other is to essentially deal only with the human and environmental consequences of ecosystem degradation. An important factor for both is the capacity of the population to adapt to either or both approaches. This, in turn, is affected by their capacity to adapt to them, which is in turn affected by their vulnerability. People in poor health are vulnerable, lack the capacity to adapt to rapid changes in environment or living conditions, and become even more vulnerable as the climate changes. This is part of the downward spiral of environmental degradation and poverty.

Although this downward spiral is a serious challenge to sustainable development, it does offer the opportunity for a rapid reversal to an upward spiral. This opportunity is dependent on reprioritizing food, shelter, and public health of the poor. This approach increases the capacity to engage in behaviors that increase the capacity to reduce ecosystem degradation. Given the increasing population and its urbanization on the 20 percent of the planet humans find habitable, this is a high priority because ecosystem degradation will spread to global systems that affect everyone. In the minds of many advocates of sustainability, this heightens the need to apply the precautionary principle to all kinds of development, no matter where it takes place. The precautionary principle requires a determination of whether the proposed development will threaten irreparable damage to the systems of nature on which future life depends. The threshold of irreparable damage may become substantially lower as cumulative effects of past environmental decisions and current population growth projections increase. It is possible that new development may be required to remedy the effects of past ecosystem degradation and to continually monitor new development and remedy efforts. Under many current economic models, this would not be profitable. Developing nations seeking traditional economic development may hesitate to use the precautionary principle for this reason, even though many of these nations have the populations most at risk of public health erosions from climate change. The downward spiral of environmental degradation and poverty may require large-scale international cooperation from developed nations and international financiers.

Environmental degradation and poverty are linked together in a downward spiral. This spiral refers to the relationship between environmental degradation and the ability of the people in that place to slow or stop the environmental degradation because of lack of income or access to other resources. As time passes, the poverty of people in a place combines with accumulating concentrations of organic and inorganic wastes. The dynamic of increasing human populations and increasing chemical potency of technological development contributes to the rate of environmental degradation and the stifling effects of poverty, hence the term *spiral*. The relationship of poverty to environment is important for equitable components of most sustainability definitions. If spiraling or other relationships create large, growing, amorphous concentrations of material that threatens systems on which all life depends, then all sustainability advocates become concerned. Some of the implications of increased scrutiny to these processes in cities is that it may require a unified and generous global effort. This can affect international political and economic relationships and redefine world order. In these controversial areas and others, new values associated with sustainability and increased environmental decision-making power are emerging as a global and domestic U.S. policy. These dynamics are examined in depth later in Chapter 3.

References

Adger, W. Neil et al., eds. 2006. *Fairness in Adaptation to Climate Change.* Cambridge, MA: MIT Press.

Edwards, Andres R. 2005. *The Sustainability Revolution: Portrait of a Paradigm Shift.* Gabriola Island, BC: New Society Publishers.

Myers, Nancy J., and Carolyn Raffensperger, eds. 2005. *Precautionary Tools for Reshaping Environmental Policy.* Cambridge, MA: MIT Press.

Pirages, Dennis, and Ken Cousins. 2005. *From Resource Scarcity to Ecological Security: Exploring New Limits to Growth.* Cambridge, MA: MIT Press.

Rappaport, Roy A. 1994. "Human Environment and the Notion of Impact." In *Who Pays the Price? The Sociocultural Context of Environmental Crisis,* ed. Barbara Rose Johnston, 157–69. Washington, DC: Island Press.

Whiteside, Kerry H. 2006. *Precautionary Politics: Principle and Practice in Confronting Environmental Risk.* Cambridge, MA: MIT Press.

Wilkinson, Richard, and Kate Pickett. 2009. *The Spirit Level: Why More Equal Societies Almost Always Do Better.* St. Albans, UK: Allen Lane.

Poverty and Exposure to Environmental Hazard: Environmental Injustice

In addition to all the other hazards of life, being poor means that the struggle for life will include an increased burden of environmental exposures. Poverty is more than a lack of income in many cultures. It is the lack of food, shelter, and fresh water. Ultimately poverty means death, and sometimes in a slow way. Poverty can mean a quality of life so low that it is barely living. Current value judgments that blame poor people for their poverty may have to be revisited under sustainability. If these value judgments affect risk assessment decisions, they may underestimate the risk that poverty poses for ecosystems and therefore for whole populations.

Reference

Cutter, Susan L. 1993. *Living with Risk: The Geography of Technological Hazards.* London, UK; New York: E. Arnold.

EQUITY AS THE CONTEXT FOR SUSTAINABILITY

Low income and political marginalization are associated with early measurements of environmental degradation everywhere in the world. Sustainability is defined as meeting the needs of the present without compromising the ability of future generations to meet their own needs. Therefore, many believe that to achieve sustainability, current generations must learn to live within the limits of remaining natural resources. This raises political, social, and economic questions related to the equitable distribution of these resources. If these natural resources, such as clean air and water, are used faster than the environment can renew, then the ability of future generations to meet their own needs will be compromised. Fairness to future generations will be difficult to secure without fairness to contemporaries, and the problems of poverty and inequitable distribution of wealth and resources must engage unresolved issues of historical injustice, as well as environmental degradation.

There are many inequitable distributions of wealth and resources and unresolved historical inequities around the world. Societies have developed in resource-rich places that place values in a context that uses them to judge people. Inherited wealth and resources, or lack thereof, is often inequitable. Individuals who receive wealth did not earn it, and individuals denied opportunities for wealth did nothing wrong to be so denied. The unfairness to future generations continues as long as the ecosystems that support them remain intact. The ecosystems are not remaining intact, however, and they affect future generations from both the wealthy and the poor. Global warming and climate change will not be eliminated by reducing the number of poor people. The top 10 percent of the wealthiest population produce almost 90 percent of greenhouse gas emissions. At this point eliminating the wealthy would not decrease global warming because the structural nature of the problem is entrenched in the historical inequities of many nations. Threats to future generations come from the current structures of these inequities and will be perpetuated by them without change. For example, many countries in Western Europe developed their colonies in places where they could extract resources with little cost to themselves but with great cost to the indigenous people. Modern multinational corporations also do the same, extracting natural resources with little cost to themselves but great cost to the developing nation. The hierarchical structure of the inequity is the same, with the same impact on the environment. Now, however, populations are expanding, resource extraction uses much more powerful technology, and the cumulative impacts of an industrialized economy are causing global warming and worldwide environmental impacts.

What Is Equity?

Equity refers to fairness. Fairness is a value perspective that can differ greatly in application between cultures and between generations. Implicit values become explicit when in conflict, which is more and more the case with environmental controversies. Groups that have privilege that is not known to them often have implicit values. Among the values in sustainability is intergenerational equity. The incorporation of equity in decision making is not new, but its incorporation in environmental decision making is recent. Even without a full and complete understanding of the contributions of other cultures and species, the environmentally destructive industrial values of the past may need to be altered if a future with the same rich diversity and life forms is to be preserved. Value changes emphasizing the gains made from inclusiveness and diversity will be articulated to promote ecological attention for the unknown content of the future. Inclusiveness demands more than the contemporary bilateral discussions between industry and traditional environmental groups, each represented by lawyers and scientists. The dialogue regarding a concept of a sustainable future , risks of technologies like nanotechnology, and how to organize natural resources and allocate them must now include many people who were external to the industrial economic values and their deficient and self-serving environmental policies. In this manner, some initial dialogues about sustainability will require capacity building and facilitation, which requires time and money.

There are several theoretical approaches to the question of fairness and to what equity can mean. Although these distinctions are theoretical, they matter because they become reflected in law and policy. The first is distributive justice. Generally, this involves distributed outcomes rather than a process for arriving at such outcomes. Equity issues in sustainability involve addressing the disproportionate public health and environmental risks borne by people of color and lower incomes by lowering risks, not shifting or equalizing current risks.

The second is procedural justice, which is a function of the manner in which a decision is made, that is, the fairness of the decision-making process, rather than its outcome. In terms of equity issues in sustainability, it matters because a community's judgment about whether a decision is just is significantly determined by the perceived fairness of procedures leading to that outcome. There is a strong value in the United States around the idea that people have a say about what personally affects them. When a violation of procedure occurs, what happens to produce a result that is distributively unjust? Although law may require compliance with the particular decision, it may be minimal. When personal consumption patterns may have to change to achieve sustainability, minimal compliance may not be enough. The new value of inclusion, however, resonates with the procedural fairness concept because participation is part of a fair procedure. This includes equal access to information, advance notice of meetings and decisions, and measurement of all environmental impacts.

A third perspective on fairness or equity is corrective justice, which is fairness in the way punishments for lawbreaking are assigned and damages are inflicted on individuals and communities. It requires the just administration of punishment and a duty to repair losses for which one is responsible. Some aspects of the compensatory aspect require fault or wrongdoing, but others impose liability regardless of fault (polluter pays principles). The equity implications for sustainability of this approach would require those that harm the environment to repair it. Currently, this is a requirement of law or policy only in exceptional cases.

The fourth theory of justice is social justice. In this theory, society uses best efforts to bring about a more just ordering of society. It strives to have members of every class have enough resources and power to live as humans. Privileged classes are accountable to the wider society for the way in which they use their advantages. The equity implications for sustainability of this approach would require those who have benefited from natural resource use to assist those who have not.

Equity influences on these approaches to sustainability include a procedural component. Sustainability includes a set of discourse and policies about the benefits and burdens of environmental decision making across race, class, culture, and gender lines. No environmental decision making can be sustainable when the health concerns of communities are ignored. The urban environment has not been treated holistically because the public health concerns of the residents are not included. This is especially true of politically and economically weak communities. Nations that treat their cities as toxic waste sites threaten all systems on which life depends, and this contradicts all three versions of sustainability.

How Does Equity Fit into Sustainability?

One version of sustainability considers mainly economic growth. This variant of sustainability calls for maximum economic growth, and the environment as a necessary precondition to this goal. This version of sustainability has dominated thinking in the United States more than other regions of the globe. According to this version, indicators of success are found in measures like gross national product or gross domestic income. These are measurements of the dollar value of certain products and transactions. These types of financial indicators are not qualified by the costs and externalities they may impose. They do not distinguish between products and services that are related to harms that should be avoided, such as cleaning up oil spills and brownfields, and those that are related to increasing the quality of life such as educational spending and preventive health care. These distinctions are important to some societies as indicators of whether progress toward sustainability is actually being attained. This has led to the creation of different measures of success such as gross national happiness, which considers the financial gains in the context of other indicators of human happiness and well-being.

This version of sustainability may seem at odds with the stated agenda of many modern-day environmentalists. It is at odds with most principles of equity because maximum economic growth for the present growing population will consume irreplaceable natural resources. Equity also asks the question of economic growth for whom. What is the environmental decision-making process by which irreplaceable natural resources are developed profitably?

A second version of sustainability is again placed in the context of economic growth, although in this context it does not seek to maximize economic growth. Economic actions that are clearly detrimental to the environment are avoided. Instead, a set of goals including security and peace, economic development, social progress, and good government are advocated. This model of development has influenced global thinking about development more than thinking within the United States. The significant contribution of sustainable development to this model is that it clearly defines all goals in terms of the limitations of our ecosystem health.

This definition often founders on implementation. A key question for implementation is whether people are part of the protected environment for business decisions. Is it necessary for science to prove causality before economic actions that are detrimental are determined? How are actions handled that may alone degrade the environment, but cumulatively become detrimental to it? Many questions about ecosystem degradation and public health remain unanswered, and the question becomes whether economic development should continue until definite answers are found. The precautionary principle could be applied here to say that if there is a risk to systems of life on which future generations will depend, then the answer to continuing development should be, no. Continuing development may simply be facilitating uncontrollable population growth, and denying development may simply decrease the quality of present public health and pose a threat to future generations. The balance between allowing economic development under this model of sustainable development and not allowing it is precarious. It is heavily influenced by the poverty of the population, the population growth, the carrying capacity of the ecosystem, and the thoroughness of the risk assessment.

A third version of sustainability is rooted in the impacts on the environment, generally including humans. In this perspective, ecosystem impacts subordinate all other considerations including human equity and economic survival. It is tied to human impacts on the land, air, and water. It requires an understanding of all our environmental impacts, including the cumulative impacts of everyday activities. This version of sustainability explicitly and implicitly challenges the existing value foundations of existing social, economic, and political institutions. Many environmentalists feel that this is the one true version of sustainability. They would argue that because all life is dependent on the conditions of the ecosystems that support life, economic and social concerns are justifiably secondary to ecosystem health. This relationship to sustainability is captured in the image of nested baskets. The environment is the largest, most comprehensive basket. Within it, all human activity is located including human economic enterprises and human communities. Economic enterprise is represented by a basket nested within our environment and its webs of life. Human individuals and their communities rest within both these two baskets, relying on each for their livelihood and support.

The environmental history and current condition of any one place is important for all sustainability policies and programs. Environmental inequities of the past and present can help explain the current ecological status and help develop meaningful environmental baselines for most communities. Most cities do not yet have the environmental information necessary to establish environmental baselines. Without environmental baseline measures, it is difficult to decide what types of behavior changes to develop, what types of rules to enforce, and how to begin a sustainable program. This is especially true in the area of cumulative impacts.

As society becomes more aware of the limitations of impacts on the environment, researchers cannot help but notice the tremendous disparities in exposure to environmental hazards domestically and internationally. The issue of sustainability now dominates discourse about environmental theories and public policy. Sustainability is harkened by some as a new emerging ecological paradigm, or way of thinking about "environment." One value of this new paradigm is justice or fairness. As attempts to implement concepts of sustainability develop, important and unavoidable questions about fairness emerge. These concerns from politically and economically disenfranchised communities raise serious public health and environmental concerns for everyone. Proponents of an equity-based sustainable vision maintain that the path to a just and sustainable environmental decision can come through the community. They maintain that the key to implementing sustainability with fairness is community involvement in decisions where they live, work, play, and worship.

EQUITY AND POVERTY

Environmental degradation and poverty are linked together. This is true in most environments, including urban environments. When people cannot meet their basic needs, they may prey on other species and damage ecosystems to survive and provide for their families. Growth of markets and expanding corporate development have had devastating consequence for both ecosystems and for the people who live closest to those systems. The empires of the 19th and 20th centuries left poverty in many conquered lands while amassing tremendous wealth in the conquerors hands. African tribes and nations lost millions of people, including skilled craftspeople and artisans, to the international slave trade. Millions of Latin American and North American people died after contact with European-borne diseases, and millions more were enslaved and their lands confiscated. The ancient empires of Asia fared no better. A contemporary consequence of this history is that many of the countries of Africa, Asia, and Latin America are struggling to meet the basic needs of their populations for food, housing, and security and increasingly to meet the crisis demands of climate change.

From the 1600s until the mid-20th century, empires colonized Africa, parts of Asia, North and South America, and parts of the Middle East. Former imperial rulers amassed wealth from the natural and human resources of colonized places. Slavery of Africans, vast land transfers in North America, and resource transfers from Asia rapidly helped to transform Western economies into consumer/industrial economies of scale, mass, and wealth; however, they left the colonies poor and indebted for generations. Even today, "developing" countries are closely identified with former colonies, and they are further saddled with debts that place human rights and dignities out of reach of many of their population. In the context of developing countries, sustainability means the ability to provide for the basic human rights and dignities of their people. In the context of developed countries, most of whom developed on the involuntary contributions of colonies, sustainability is focused on issues of sustainable production and consumption, not basic human rights. Both developing and developed countries, however, are now so interconnected by geographic realities, such as watersheds, air sheds, trade, communication, finance, and other structures, that ideas like territory, nationality, sovereignty, race, and even private property must concede our mutually connected destinies.

Sustainability includes all people as a moral and ethical commitment. An ecumenical group of spiritual organizations has also joined to advocate for the protection of the environment and provision of assistance to the poor. They drafted the "Earth Charter," a statement of principles of responsibilities based on faith. This group is an independent global organization representing more than 2,500 organizations, 400 cities and towns, global agencies like UNESCO (United Nations Educational, Social, and Cultural Organization) and IUCN(International Union for Conservation of Nature) and individuals. The charter has four basic principles subdivided into 16smaller tenets. The four principles are respect and care for the community of life; ecological integrity; social and economic justice, including the eradication of poverty; and democracy, nonviolence, and peace.

The Earth Charter also includes all of us as a practical matter because human societies are now so interconnected by communication, transportation, trade, and other entities. Yet inclusiveness of all is notably lacking in U.S. public policy on sustainability. Urban areas, low-income communities, people of color and their communities, women, and children often are excluded from participation in environmental decisions that affect their health and well-being. In the United States, the environmental justice movement has demanded inclusion of these interests and provided leadership for inclusive public policy. This inclusiveness is an essential element of sustainability. Globally, the Civil Society movement has taken on the work of strengthening commitments to inclusiveness and equity by using nongovernmental partners to mobilize and increase capacity to participate in environmental decision making.

Sustainability in this area is a process as well as a goal. Inclusive public participation in environmental decision making facilitates enforcement and implementation of solutions. A decision that is widely accepted as fair is likely to be self-implementing. Sustainable decision making must be transparent, accessible, and accountable to all interested stakeholders. Any environmental decision will touch the interests and concerns of many people and beings with which we currently share this planet, as well as future people and beings. Inclusive processes must strive to incorporate all interested parties or their representatives in problem identification and resolution. Especially where information is lacking and values are in dispute, a fully inclusive process ensures that values important to all stakeholders will be heard and may guide decision making in the face of uncertainty.

Transformation of the Slums of Mumbai and Advocacy by Slum Dwellers

The case of the slum dwellers and rag pickers of Mumbai, India, has grown into an international context. The city grew around the slum, which became a transportation hub, and the government granted squatters limited property rights in their huts, as well as providing public services like water and electricity.

The population of India is about 1 billion people, and the urban population is about 285.4 million people. Areas that are without adequate municipal services, such as fresh water and housing, are called slums. There are approximately 100 million slum residents in India. In many nations, the majority of poverty exists in rural populations, but with increasing urbanization comes increasing urban poverty, which greatly threatens existing ecosystems. The city of Mumbai is both the commercial center and slum dweller center for India. Mumbai has a population of 12 million people, of whom about 6 million are considered slum dwellers.

The dominant political culture does not value slum dwellers and marginalizes them politically and economically. As a result, there is little fresh water, adequate shelter, or roads. Slum dwellers live on about 16 percent of the land mass of the city in densely concentrated conditions.

Partly as a result of the extensive oppression, a civil society called the Society for the Promotion of Area Resource Centers formed as a nongovernmental organization in 1984. Its main core groups were known as "women pavement dwellers." These were women whose homes had been repeatedly demolished and who were made homeless. This eventually evolved into Mahila Milan (Women Together), incorporating other slum dweller community organizations. These groups have spread to other nations in Africa and Asia where cities are dominated by large and growing slums. The core organizing principle of Mahila Milan is the recognition of the role of women in the family and in affecting change in impoverished circumstances. The organization engages in all sorts of training, credit improvement, and advocacy activities.

References

Collier, Paul. 2009. *The Bottom Billion: Why the Poorest Countries Are Failing and What Can Be Done about It.* London, UK: Oxford Press.

Sachs, Jeffery. 2006. *The End of Poverty: Economic Possibilities for Our Time.* New York: Penguin Press.

Values: Historical Human Values

The environmental interaction of humans through history evidences a range of approaches and values to the environment and to each other. Many would judge these actions harshly by the standards of today. Human history is filled with acts of wanton and unnecessary destruction, senseless acts of greed, and sad oppression of truth and knowledge. These acts evidence values. When human populations were low, environmental impacts did not challenge the carrying capacity of their ecological systems. When these very acts are reexamined through a lens that is more knowledgeable, and when human populations are challenging the ecological carrying capacity, the judgment can be harsh. As sustainability examines the systems on which all life depends, future historians may judge our ignorant environmental actions of today just as harshly. Sustainability advocates often uncover uncomfortable evidence of currently rejected values, such as human servitude. These old values, whether rejected today or not, are the historical foundation for human interaction with the environment.

References

Ackerman, Frank, and Lisa Heinzerling. 2004. *Priceless: On Knowing the Price of Everything and the Value of Nothing.* New York: New Press.

Atran, Scott, and Douglas Medin. 2008. *The Native Mind and the Cultural Construction of Nature.* Cambridge, MA: MIT Press.

Geisler, Charles, and Gail Daneker, eds. 1997. *Property and Values: Alternatives to Public and Private Ownership.* Washington, DC: Island Press.

O'Neill, John, Alan Holland, and Andrew Light. 2008. *Environmental Values.* Andover, UK: Routledge.

Paavola, Jouni, and Ian Lowe. 2005. *Environmental Values in a Globalizing World: Nature, Justice, and Governance.* Andover, UK: Routledge

Zazueta, Aaron Eduardo. 1998. *Policy Hits the Ground: Participation and Equity in Environmental Policy Making.* Washington, DC: World Resources Institute.

Civil Societies and Sustainability

Civil society as a movement recognizes the roles of nonprofit organizations. There can be many different types of nonprofit organizations, and they are given special tax treatment in the United States. They do not necessarily have to be environmental organizations, but many of those organizations take a lead role in advocating for sustainability. Civil societies around the world are those organizations that are apart from business or government. They are often nongovernmental organizations (NGOs). The range of civil societies is large and informal. Community organizations of all types—religious organizations, professional organizations, labor unions, and school organizations—are civil societies. The idea behind their inclusion is that they provide a strong voice of the people in democratic nations.

Some areas around the edges of a civil society are profit-motivated news media. They are corporations interested in making a profit, so their bias in contributing to the public discourse is often influenced by that perspective. Some organizations are motivated solely by exclusion of others. Some groups exclude people on the basis of race, gender, or religion. Some groups advocate violence against these groups. Other civil society groups may advocate for peace and inclusion. The idea of civil societies is not value free or value neutral, especially for purposes of sustainability. Another issue is the role of state support for civil societies. Many countries give special tax treatment to them, and many Western European countries give them direct subsidies. Some countries give preferential subsidies to civil societies that share values that may be counter to sustainable development. Many environmental civil societies, for example, do not receive the same subsidies as do the primary religions of a particular nation.

The explicit value of civil societies is that they are inclusive, which is a strong value in the equity component of sustainability. Civil societies are also valuable for purposes of sustainable development because they can bring expertise, knowledge, and resources to aspects of sustainable development such as risk assessment. As such, they are being increasingly included in worldwide efforts aimed at sustainable development.

The World Bank began working with NGOs on environmental issues in the 1970s, and this was their first experience with civil societies. Now the World Bank explicitly consults and collaborates with thousands of civil societies. The United Nations commissioned a special study on the role of civil societies. As the lead author of the report notes:

> The rise of civil society is indeed one of the landmark events of our times. Global governance is no longer the sole domain of Governments. The growing participation and influence of non-State actors is enhancing democracy and reshaping multilateralism. Civil society organizations are also the prime movers of some of the most innovative initiatives to deal with emerging global threats.
>
> Given this reality, the Panel believes that constructively engaging with civil society is a necessity for the United Nations, not an option. This engagement is essential to enable the Organization to better identify global priorities and to mobilize all resources to deal with the task at hand. We also see this opening up of the United Nations to a plurality of constituencies and actors not as a threat to Governments, but as a powerful way to reinvigorate the intergovernmental process itself.
>
> The world stands today at a very delicate juncture. The United Nations needs the support of civil society more than ever before. But it will not get that support unless it is seen as championing reforms in global governance that civil society is calling for—and which are echoed in our report.
>
> Kofi A. Annan, June 11, 2004

The United Nations especially recognizes the role of civil societies in sustainable development. This is a delicate issue for the United Nations because member states do not want their civil society organizations to subvert their power, as noted previously.

The civil society movement holds great potential for the equitable component of sustainability. The major world organizations advocating for sustainable development recognize the need to bring in all groups of people, whether or not the government includes them. It is necessary to truly protect and evaluate ecosystem viability knowledge from these groups. Without this knowledge of ecosystem risks, other aspects of sustainability become difficult to implement. For example, the application of the precautionary principle, which determines if there is a irreparable threat to the systems of nature on which future life depends, is stymied without accurate and complete knowledge of the ecosystem. In many nations, however, there is resistance from incorporating the civil societies of that place. The values of civil society organizations range over a wide spectrum and can change over time. They can be influenced by the economic and political power of the nation, or the nation can be at odds with the values of the civil society. International organizations like the ones discussed are organized around nations and as such are very careful when they engage civil societies. They do not want to threaten their member nations with inclusion of civil societies. Other international organizations in the private sector are often sought to finance or manage large developments with substantial environmental and economic impact. Their perspective on civil societies is one of less engagement. They do not want to interfere with their potential for profit or engage in the political affairs of the nation. As such their environmental impacts can be unchecked by civil societies, and these societies may be the only ones knowledgeable of the impacts. This develops the dynamic of civil societies in controversial environmental projects going to international bodies and circumventing their governments. To the extent that processes are in place that includes civil societies, the equitable component of sustainability increases.

Reference

Annan, Kofi A. 2004. We the Peoples: Civil Society, the United Nations and Global Governance: Report of the Panel of Eminent Persons on United Nations–Civil Society Relations, p. 2, A/58/817, June 11, 2004.

Betsill, Michele M, and Felix Corell, eds. 2007. *NGO Diplomacy: The Influence of Nongovernmental Organizations in International Environmental Negotiations.* Cambridge, MA: MIT Press.

Sustainability as an Equitable Process

When environmental decisions are in gridlock, or when the decision-making mechanisms of that society fail to make sound environmental decisions, values become apparent. Actions determine values. Ethics is a way to view human behavior as right or wrong. What happens when humans act unethically or the ethics of one group conflict with another? What happens to an ecosystem over time when environmental decision making and policy are designed for economic development and not environmental protection? These are environmental areas where values are now in gridlock. Yet, environmental degradation continues in many places. In terms of environmental impacts, the decision to go with the status quo, the usual way of doing, or business as usual, only makes the environmental impacts worse. Increasing knowledge, however, has alerted environmental decision makers and new active stakeholders like communities to the unexamined values that occur as a default when values are in conflict. Sustainability proponents push open this area of unexamined values and go far beyond a professional or religious base of ethics.

CHALLENGES TO INCLUSION

Many challenges interfere with inclusive participation of all environmental stakeholders. Stakeholders may have very different capacities for participation based on differences in education, access to technology, language skills, cultural norms, and special sensitivities and vulnerabilities. Norms and expectations surrounding the process may affect participation including norms around race, gender, age, and class. Other policy issues such as national security issues and confidential business information may affect access to important information. Unacknowledged biases such as privilege and deeply held assumptions about property may be the most difficult challenges to inclusive dialogue aimed at achieving sustainable goals.

Concepts of fairness and equity can be complex when discussed alone. When equity concepts are applied to emerging concepts of sustainability, this complexity can be overwhelming. Equity issues in sustainability usually concern people. People live in cities now more than ever. At no time in the history of the planet has any one species dominated the environment as much as ours. By the year 2030, 60 percent of the world's population will be living in urban settlements—cities and towns. The majority of people will be living in less developed countries, and 84 percent of people living in developed countries will be living in cities. Because of global warming, climate change, and rising ocean levels, many cities could suffer more environmental impacts from natural disasters like flooding, high winds, fires, and droughts. Many cities, especially in the United States, are along ocean coasts and will be affected by rising ocean levels and different weather changes. Urban areas are vulnerable to these natural disasters. Natural disasters can spread dangerous pollution deeper and further into a given ecosystem. Incorporating environmental issues of urban areas for sustainability may require policies that address urban vulnerability to natural disasters. An example of such a policy would be an effective alert and evacuation plan. Some countries with experience with hurricanes and the natural disasters around them have evacuation routes designated with safe houses and with provisions for fire, police, and emergency personnel. Other nations without experience and planning with natural disasters may pose challenges for urban environmental planning and for sustainability.

In the United States, the land with the most intense human footprint, the weakest enforcement of environmental laws, and the most need for sustainability is decidedly urban. Urban areas are much more complex than parks or other areas with low population.

Environmental policy in urban areas across the United States is relatively new. The U.S. Environmental Protection Agency (EPA) was formed in 1971, about the same time as many state environmental agencies. U.S. urban areas had at least a century of unrestrained industrialization, with no environmental regulation and often no land use control whatsoever. Therefore, all the environmental consequences of these acts were "unintended." U.S. environmental movements focused on unpopulated areas,

Year	Urban Population
1950	736,796
1955	854,955
1960	996,298
1965	1,160,982
1970	1,331,783
1975	1,518,520
1980	1,740,551
1985	1,988,195
1990	2,274,554
1995	2,557,386
2000	2,853,909
2005	3,164,635
2010	3,494,607
2015	3,844,664
2020	4,209,669
2025	4,584,233
2030	4,965,081
2035	5,341,341
2040	5,708,869
2045	6,063,186
2050	6,398,291

Figure 3.3 • The number of megacities that have 10 million or more residents is increasing worldwide: 1950: 4, 1980: 28, 2002: 39, 2015: 59. According to the estimation of the UN concerning the number of megacities in 2015, Bombay (22.6 mill. inhabitants), Dhaka (22.8), Sao Paulo (21.2), Delhi (20.9), and Mexico City (20.4) will be five of the worldwide six biggest megacities, each with much more than 20 million inhabitants. Population Division of the Department of Economic and Social Affairs of the United Nations Secretariat.

not cities. U.S. environmental movements did not consider public health as a primary focus, but emphasized conservation and preservation of nature and biodiversity. Cities were also the dynamic melting pot of new immigrants, and three waves of African Americans migrating north after the Civil War. According to one well-known conservative scholar,

Nathan Glazer, African Americans have not melted into U.S. society. Formerly, a proponent of cultural assimilation and against affirmative action, Professor Glazer says he was wrong about African Americans. This group faced substantial discrimination in housing, employment, education, and municipal services. As industry and technology rapidly expanded in the city, so, too, did this population. As waste from these industries increased and accumulated over time in the cities, so, too, did the exposure from these wastes faced by immigrants, African Americans, and low-income people generally. These populations face tremendous displacement pressure, but African Americans and other people of color also face difficult challenges in obtaining new housing within the same community (or elsewhere) after displacement. For example, when these populations are displaced, they often face paying a disproportionately high percentage of income for housing, as well as suffer from the loss of important and intangible community culture. Market forces and social attitudes underlie displacement of residents and provide an awkward context for equity components of sustainability.

References

Glazer, Nathan. 1998. *We Are All Multiculturalists Now*. Cambridge, MA: Harvard University Press.

Wilkinson, Richard, and Kate Pickett. 2009. *The Spirit Level: Why More Equal Societies Almost Always Do Better*. St. Albans, UK: Allen Lane.

The Role of Faith: "Earth Charter"

Sustainability includes all people as a moral and
ethical commitment.

The Earth Charter, 2000

The Earth Charter is described on the Earth Charter Initiative Web site as a widely recognized, global consensus statement on ethics and values for a sustainable future. It is a declaration of fundamental values and principles considered necessary for building a sustainable global society in the 21st century. The Earth Charter was not successful in gaining support and adoption at the 1992 Rio Earth Summit. Two years later, however, the time was right and in 1994, the movement to draft and implement the Earth Charter was restarted as a civil society initiative. The renewed efforts were the work of Maurice Strong, chairman of the Earth Summit, with the help of the Earth Council, and Mikhail Gorbachev, with the support of Green Cross. The government of The Netherlands also acted as a cornerstone of the renewed Earth Charter movement. The drafting of the text was overseen by the independent Earth Charter Commission, and the Commission continues to serve as the steward of the Earth Charter text.

The Earth Charter was completed in March 2000 and launched at The Peace Palace in The Hague. Since its launch, The charter has been formally endorsed by thousands of organizations representing millions of people around the world. By 2005, the Earth Charter had become widely recognized as a global consensus statement on the meaning of sustainability, the challenge and vision of sustainable development, and the principles by which sustainable development is to be achieved.

Beginning in 2007, national governments began to make even stronger, more formal commitments to the Earth Charter. Although not endorsed by any international governmental organization, the Earth Charter has a strong impact on global guidelines for sustainability and development goals. The Earth Charter continues to have increasing international impact as a source of inspiration for action, an educational framework, and a reference document for the development of policy, legislation, and international standards and agreements.

The Earth Charter lives up to its name, having been created by a large global consultation process. The charter encourages the global community to help create a global partnership at a critical time in world development. The purpose of the Earth Charter is to inspire in all people a sense of interconnectedness and interdependence, as well as a shared responsibility for the human family and the larger living world. The Earth Charter urges environmental responsibility, peaceful coexistence, and respect for life, democracy, and justice to achieve its goals. The Earth Charter's ethical vision proposes that environmental protection, human rights, equitable human development, and peace are interdependent and indivisible. It provides a new framework for thinking about and addressing these issues. The Earth Charter is organized into 16 general headings, each covering a general principle. *See also* **Appendix C: The Earth Charter.**

References

Benstein, Jeremy. 2006. *The Way into Judaism and the Environment.* Woodstock, VT: Jewish Light Publications.

Chapple, Christopher Key, and Mary Evelyn Tucker, eds. 2000. *Hinduism and Ecology: The Intersection of Earth, Sky, and Water.* Cambridge, MA: Center for the Study of World Religions Harvard University Press.

Folz, Richard C. et al., eds. 2003. *Islam and Ecology: A Bestowed Trust.* Cambridge, MA: Center for the Study of World Religions Harvard University Press.

Gottlieb, Roger S. 2004. *This Sacred Earth: Religion, Nature, Environment.* Andover, UK: Routledge.

Mission, C. F. 2005. *Sharing God's Planet: A Christian Vision for a Sustainable Future.* London, UK: Church House Publishing.

Northcott, Michael S. 1996. *The Environment and Christian Ethics.* Cambridge, UK: Cambridge University Press.

Voluntary Simplicity

To the extent that our ecological footprints are related to our capacity and desire to consume goods and services that pertain to our comfort, redefining happiness and consumption related to getting it may be part of sustainability. A movement to examine and redefine our consumer assumptions and patterns has developed with the express purpose of trimming our voracious appetites in the best interests of our ecology. This movement exists in tension with economic policies that encourage consumption for the sake of economic revitalization.

Individuals who are interested in reducing the impact of their activities on their environment can take personal action to reduce their environmental footprint. Many individual steps are easily incorporated into daily life activities. The impact of these changes may seem small until they are adopted by a significant number of individuals, even if they are less than a majority. The aggregate effect of individual changes can affect changes at a much larger level. Some authors have argued that a "tipping point" for human behavior change exists at 10 percent. Conversely, individuals confronted with the necessity for wide-scale change may feel that their individual efforts lack significance. This psychological state of mind is called lack of self-efficacy.

Many people motivated to change their individual lifestyles to reduce their impacts on our ecosystems have formed a movement toward voluntary simplicity. As a contemporary movement connected to sustainability, this movement deemphasizes consumer behaviors that consume energy and create waste. It has many other aspects connected to spiritual practices and economic pragmatism.

This movement is voluntary. There is no enforcement mechanism and no certification organization. Nevertheless, a central feature of this diverse movement is the idea that simple living can be fulfilling, rewarding, even rich, without excess and harm to our environment. Its value to sustainability is that it is inclusive of all people. It implies that approaches to sustainability do not need to be scientifically complicated.

References

Dauvergne, Peter. 2008. *The Shadows of Consumption: Consequences for the Global Environment.* Cambridge, MA: MIT Press.

Princen, Thomas, Michael F. Maniates, and Ken Conca, eds. 2002. *Confronting Consumption.* Cambridge, MA: MIT Press.

HUMAN HEALTH AND THE ENVIRONMENT

Human health is directly affected by poverty. Access to medical care, nourishing food, clean water, and education is reduced by poverty. Poverty is also contextual. Poverty in a rich nation may mean low income only. Poverty in a poor nation may mean death in early childhood. The health effects of poverty do not limit themselves to the population of poor people. Diseases encouraged by poverty and environmental degradation, such as tuberculosis, spread to all parts of the population. The Millennium Assessment closely analyzed the future health of poor people under different scenarios of sustainability.

Human health of poor people improves in the future under most of the sustainability scenarios. The number of children affected by undernourishment is reduced. Rates of HIV/AIDS, malaria, and tuberculosis also decrease. Improved public health measures limit the impact of new diseases such as SARS (severe acute respiratory syndrome), a recently identified acute respiratory syndrome caused by the SARS coronavirus. This virus is linked to environmental degradation because it has crossed the species barrier from animals to humans. Signs and symptoms are similar to flu at the outset but develop into to pneumonia-like symptoms. Most infected patients recover if they have access to medical care, which poor people lack. It can be fatal and is easily transmitted, especially in crowded conditions. Under the Order of Strength scenario developed by the United Nations, health and social conditions for the North and South could diverge, causing a negative spiral of poverty, declining health, and degraded ecosystems in developing countries. Rich nations close down their borders and externalize global threats to poor countries, causing an increase in the systems of nature on which future life depends. This regionalization is considered a problem when dealing with world poverty. The Millennium Assessment attempts to deal with different Order of Strength scenarios.

Each scenario yields a different outcome of gains, losses, and vulnerabilities to human health in different regions and poverty populations. For example, globally integrated approaches that focus on technology and property rights for ecosystem services generally improve human well-being in terms of health, security, social relations, and material needs. This aspect of human health is considered a mosaic. Using technology for ecosystem services is called a technogarden. Globally integrated approaches to human well-being and health that use the technology for ecosystem assessment and improvement can reduce the consequences of poverty. *See also* **Volume 3, Chapter 4: Poverty and the Environment; Volume 3, Chapter 5: Science for Policy.**

Reference

Robson, Mark G., and William E. Toscano. 2007. *Risk Assessment for Environmental Health*. Hoboken, NJ: Jossey-Bass.

Poverty and Adverse Impacts: Their Relationship to Fairness and Sustainability

Because of the limitations and controversies surrounding adverse impacts, their relationship to equitable aspects of sustainability is important. One of the major limitations is that adverse impacts are those that are proven by science to be medically harmful. Adverse impacts may have to be redefined to mean impacts that degrade natural systems on which all life depends, a principle of sustainability. Many health conditions that are not considered adverse do erode the quality of life, decrease worker productivity, and are attributed to accumulating pollution. Asthma is one example. Asthma is difficulty breathing. It has adverse respiratory health impacts. The definition of adverse respiratory health effects is "medically significant physiologic or pathologic changes generally evidenced by one or more of the following: (1) interference with the normal activity of the affected person or persons, (2) episodic respiratory illness, (3) incapacitating illness, (4) permanent respiratory injury, and/or (5) progressive respiratory dysfunction." Air quality and an expanded application and definition of adverse impacts may reveal more significant public health impacts than are now known.

Whether in a developed country or in a developing country, poverty and the inability to meet basic needs for food, shelter, and care make some human communities even more vulnerable to environmentally degraded conditions. Low income and political marginalization are associated with early measurements of environmental degradation. People faced with exposure and hunger will contribute to environmental degradation to meet basic life needs. People driven by insecurity as to basic living conditions are likely to accept almost any employment regardless of consequences to human and environmental needs.

Most residents of environmentally degraded communities are generally poorer in terms of income and access to resources. In the United States and across the planet, unemployment rates can be persistently high in cities over a long time. Engagement with these communities for purposes of inclusion under sustainability will stop any dangerous environmental degradation and deal with empowerment of the residents to understand and control their environment.

Job Blackmail

The ability of employers to control the behavior of their employees outside of their work activities is controversial and a dynamic parameter of liberty values. In most Western management models, the managers are required to have professional, 24-hour devotion to the business. Workers must do the required parts of their job when at work. Some governmental protections exist against employers exercising coercive force on employees, such as threatening them with job loss unless they vote a certain way, but the common expectation is that an employee would not take a stance against the profitability of her employer. This is one way the value of profit is instituted as a proxy for social good In the industrial age of natural resource exploitation through mining, logging, and grazing, many communities grow around the site and then around manufacturing plants. Logging towns, gold rush towns, oil towns, and others all suffer when the natural resource becomes depleted and is not renewable fast enough for the current generation to make a profit. Towns that are reliant on a single company base are called company towns. This can be any large entity that dominates a community, such as a university or government agency like the Fish and Wildlife Service. Especially in industrial manufacturing towns, these communities bear a disproportionate amount of environmental pollution and subsequent degradation. Even in logging towns, the chemicals used to keep the saws running still affect the groundwater once the mills are closed. Among blue collar workers, the fear of losing one's job because of environmental regulations is a major concern. Some affected workers feel strongly that it is their right to accept more hazardous and environmentally degrading risk. Under a regime of sustainability, any such actions that could cause irreparable damage to the ecology of a place would not be permitted.

The effects on labor of decision making under sustainability are unknown. Labor will be a voice in inclusionary dialogues; the fear of job loss as a result of environmental regulation will again be voiced. For the most part, most environmental regulations do not throttle industry. Many other factors affect labor market demand. In 2008, energy costs exceeded labor costs for most U.S. industries for the first time. See discussion in Volume 2.

Sustainability advocates understand how environmental issues are related to other development issues. Although it may be true that economic market forces move those without income to less desirable and more polluted areas, over time these environments suffer cumulative effects of pollution that affect the entire ecosystem in terms of natural systems on which all life depends. Intervening in natural systems only in affluent neighborhoods is like putting a finger in a dyke to stop a flood. It works for a short time for a small number of people. Systemic approaches advocated by many sustainability theorists attack the root cause of the worst pollution first, which often brings them in contact with low-income populations.

Reference

Kazis, Richard. 1982. *Fear at Work : Job Blackmail, Labor, and the Environment*. New York: Pilgrim Press.

Human Poverty Index

Poverty means that income is so low it is difficult to meet basic food, shelter, and clothing needs. Generally, 5 to 15 percent of the U.S. population is relatively poor compared to other U.S. cities. The depth and size or global urban poverty are much bigger. Desperate communities accept almost any risk for a job. They become focal points for environmentally degrading and locally unwanted land uses. Waste sites and waste transfer stations, metal fabrication plants, slaughterhouses, tanneries, incinerators, bus depots, and auto body repair shops are generally undesirable land uses associated with poor communities. Generally, the lower one's income, the higher the risk of exposure to environmental and occupational hazards at home, school, and work. In addition, the lower one's income, generally, the greater the risk of diseases caused or made worse by environmental factors. The level of municipal environmental service provision in the traditional categories of sanitation and garbage pickup, public works projects to mitigate environmental impacts, and transportation is less and of inferior quality for poor people. Poor neighborhoods generally receive fewer parks and open outdoor recreation areas.

How the Human Poverty Index (HPI) Is Calculated

HPI is assessed through three components: longevity, knowledge, and standard of living. The HPI is divided into two indicators, HPI-1 and HPI-2. HPI-1 is used to evaluate deprivation in most countries of the world, whereas HPI-2 is used to calculate deprivation in selected OECD (Organization for Economic Co-operation and Development) countries. HPI-2 includes a fourth component of social inclusion.

The elements of HPI-1 are calculated in the following way. To determine longevity, the HDI uses the definition "vulnerability to death at a relatively early age, as measured by the probability at birth of not surviving to age 40." The indicator of knowledge is measured through the adult illiteracy rate. The last indicator of a decent standard of living is calculated through "the unweighted average of the percent of the population not using an improved water source and the percentage of the children under weight-for-age." The formula used to calculate HPI-1 is: HPI-1 = $[1/3\,(P_1^{\alpha} + P_2^{\alpha} + P_3^{\alpha})]^{1/\alpha}$. The formula is calculated with each element described above represented by $P_{1,2,3}$ and $\alpha = 3$.

The formula for HPI-2 is almost identical to that of HPI-1; however, the elements are determined in a different way. The first indicator, longevity, is calculated by the "vulnerability to death at a relatively early age, as measured by the probability at birth of not surviving to age 60." Knowledge is measured by the percentage of adults (ages 16–65) lacking functional literacy skills. The decent standard of living is measured by the percentage of people living below the income poverty line (or 50% of the median adjusted household disposable income). The fourth element, social inclusion, is determined by the rate of long-term unemployment lasting 12 months or more. The formula used to calculate HPI-2 is: HPI-2 = $[1/4\,(P_1^{\alpha} + P_2^{\alpha} + P_3^{\alpha} + P_4^{\alpha})]^{1/\alpha}$. The formula is calculated with each element described above represented by $P_{1,2,3,4}$ and $\alpha = 3$.

The Human Poverty Index (HPI) is a measure of the standard of living in a given country. It was developed by the United Nations and is often used in conjunction with the Human Development Index. The HPI indicates to what extent people in a country are not benefiting from development.

POVERTY: THE INTERNATIONAL CONTEXT

Debt Forgiveness and Colonialism

Developed and developing nations often share a common history of colonialism. Developed nations such as Britain, Spain, Portugal, Austria, and others would "colonialize" other nations. Developed nations would control the political structure and economic activity of the colonized nation. Sometimes the domination would be maintained by military force and suppression of the indigenous population. Sometimes the major social institutions of the developing nation, such as religion, education, and family structure, would be dominated by the colonizing nation. In most cases, the developed country was using the developing nation to extract resources for international trade and for their own profit. Colonialization occurred all over the world in every continent. Its environmental consequences last until the present day.

One example of the role of colonization, environmental degradation, and poverty is the continent of Africa. Africa is a large continent with enormous biodiversity, huge rural poverty, a past and present of colonization, and continuing environmental degradation. Africa was used by developing nations for slaves and natural resources. West European nations sought humans for slaves and other items such as manioc, palm oil, and ivory. They restructured the way Africans traditionally farmed and otherwise used their land. European sport hunters established game preserves without any regard to the subsistence requirements of African. This forced African farmers to overuse other land in less productive areas. Modern rural poverty in Africa is a result of some of these practices. European traders often required colonies to grow the crops they wanted, which were often nonindigenous and displaced indigenous crops. The emphasis on monocultural cash crops still exists today as now free African nations try to grow these crops to pay off their international debts.

European colonizers often pursued environmentally degrading practices such as mining, cash crop farming, logging, and grazing. They did so to a much greater extent than they did in their home countries. In postcolonial Africa these practices were abandoned, and many of their more severe environmental consequences have ended. One of the most severe consequences of global warming and climate change, however, is the desertification of sub-Saharan Africa.

One of the remnants of colonialism in Africa is the foothold it gave private companies. These companies, now multinationals, continue to exploit environmental resources in Africa such as diamonds, oil, forests, animals, gold, and other monocultural cash crops. They often have international protection for their activities. African nations with ingrained rural poverty have engaged in programs with the World Bank and International Monetary Fund (IMF) that allow private international corporations to continue colonial practices because these agreements do not allow the African nations to regulate the activities of the multinational corporations or require them to mitigate environmental damages.

The present-day environmental context of Africa is also one of environmental injustice. As the amount of toxic waste has increased and developing nations become aware of its dangers in their home countries, they have begun to dump it in Africa. This has occurred in Egypt, Nigeria, the Republic of Benin, and Guinea Bissau. Some countries allow this to occur to get money to pay off international debts.

References

Agarwal, Anil, and Sunita Narain. 1991. *Global Warming in an Unequal World: A Case of Environmental Colonialism.* New Delhi, India: Centre for Science and Environment.

Dowies, Mark. 2009. *Conservation Refugees: The Hundred-Year Conflict between Global Conservation and Native Peoples.* Cambridge, MA: MIT Press.

Grove, Richard. 1998. *Ecology, Climate and Empire: Colonialism and Global Environmental History, 1400–1940.* Isle of Harris, UK: White Horse Press.

Rappaport, Roy A. 1994. "Human Environment and the Notion of Impact." In *Who Pays the Price? The Sociocultural Context of Environmental Crisis,* ed. Barbara Rose Johnston. 157–69. Washington, DC: Island Press.

Debt Relief: Postcolonial Responsibilities and Sustainability

The debts of postcolonial nations are very high and their ability to pay them continues to decrease. There is great concern that postcolonial developing nations will suffer the most from climate changes and become even less able to adapt to them. The pressure on the IMF and the United Nations to meet the Millennium Development Goals (MDGs) by 2015 has caused a realignment of policy to engage sustainable development. The IMF wants nations to develop their own policies to meet the MDGs and to aim them at reducing poverty. By reducing poverty and becoming more sustainable, they can pay more of their debts and require less international aid.

For postcolonial nations to receive aid, they must begin to integrate economic development into environmental sustainability, which can mean a renewed focus on natural resource management. This will require technology transfer to developing nations if their economic development models are to be different from the industrial economic models of developed nations.

A major source of tension is that developed nations used industrialized methods to develop their economies. These methods dramatically increased both greenhouse gas emissions and the quality of life. Postcolonial developing nations want to increase their quality of life, too, and they do not want to be forced into colonial status. This is known as the North–South conflict. This means that developed nations mainly in the Northern Hemisphere are imposing different standards and structures for developing nations of the Southern Hemisphere. The greenhouse gas emissions of the northern countries, such as the United States and Western Europe, are major contributors to climate changes that affect developing countries. Northern nations continue to be large greenhouse gas emitters, and some, such as the United States, have not signed international treaties such as the Kyoto Protocol. If developing nations use the same industrial methods to improve their quality of life as developed nations, however, climate changes could be much more severe. Add to this complex problem the growing populations in developing nations and their entrenched poverty. Essentially many nations are being asked to develop in ways they cannot afford. This is why technology transfer is so essential.

The Global Monitoring Report is an annual report that evaluates success or failures in achieving the necessary MDGs to reduce postcolonial debts in ways that are sustainable. *See also* **Volume 3, Chapter 4: Poverty and the Environment.**

Reference

World Bank. 2008. *The Global Monitoring Report.* Washington, DC: World Bank.

Millennium Development Goals, Targeted Investment Areas

In 2000, the General Assembly of the United Nations adopted a declaration of goals for development to end poverty by 2015.

The Millennium Campaign adopted eight goals:

1. End hunger

2. Universal primary education

3. Gender equity

4. Child health

5. Maternal health

6. Combat HIV/AIDS

7. Environmental sustainability

8. Global partnership

Some of the MDGs target investment into some of the most impoverished areas of the world. Although the rate of population and poverty is increasing in urban areas, it is endemic to rural areas. One reason it is increasing in cities is that the rural areas are so impoverished that people leave them to pursue opportunities in the cities.

The most outstanding trend of global poverty is its concentration in the rural areas of developing countries. The more remote an area, the more likely that it is an area of intense poverty. Remoteness is generally a function of connection to transportation. Worldwide, about 75 percent of the people living in absolute poverty live in rural areas. Even in areas that are urbanized, most poor people are in the rural areas. For example, in Latin America and the Caribbean, about 75 percent of the population lives in the cities, but 40 percent of the poor people live in the rural areas.

Another characteristic of world poverty is that rural poverty levels are always higher than nationwide levels of average poverty. In most areas of the world, MDGs have not reached the rural poor.

The challenge to rural poverty is to grow enough food to survive. This requires water and healthy soils. It also requires roads and other transportation linkages. In places like Africa, road density remains very low so that the benefits of the Green Revolution in agricultural productivity never occurred. This causes necessary items like fertilizers and pesticides to cost a prohibitive amount. Communication linkages to best practices for anything agricultural require education and the ability to communicate over long distances. Without education and communication linkages, the knowledge that could assist subsistence and economically productive agricultural rural communities fails to arrive.

In terms of sustainability, rural poverty poses many challenges. Climate changes and feeding an increasing world population stress a rural population already below subsistence levels. *See also* **Appendix B: The Millennium Development Goals.**

Environmental Refugees: Lives Disrupted by Environmental Degradation

Environmental refugees are often also referred to as climate refugees. They are people displaced by climate change and environmental disasters. The number of environmental refugees will continue to increase with the significantly higher number of environmental disasters that have inflicted the world in recent years. Environmental refugees have been described as those people who have been forced to leave their traditional habitat, temporarily or permanently, because of a marked environmental disruption (natural and/or triggered by people) that jeopardized their existence and/or seriously affected the quality of their lives. It is important to note that environmental refugees are not recognized by the United Nations, and international norms have yet to change to include these victims of global environmental change.

Katrina: Refugees in Your Own Country

Hurricane Katrina slammed into the southern Louisiana coast and New Orleans on August 28, 2005. Eventually about 1,800 people died. Most were African American. Many more were physically injured and many more than that were emotionally scarred. The damage to property was enormous. The damage to the environment was even larger because the numerous chemical manufacturing plants and oil refineries simply let their wastes, products, and plants enter the tidal surges. Municipal services like police, fire, sanitation, and emergency services vanished. During and after the hurricane, it was a place too dangerous to live in. Many people fled and they were called refugees. Again, most were African American.

They were environmental refugees, but controversy occurred because of the slow response time by the federal government and the treatment of people during and after the hurricane. Many people compared the federal response to Hurricane Katrina to other hurricanes, and attributed its slow response and aftercare to racism. Never before had victims of environmental catastrophes been called refugees. They were not always welcome. The victims of Hurricane Katrina dispersed over 28 states. About 240,000 went to Texas, 60,000 to Arkansas, and about 50,000 stayed in other parts of Louisiana. As global warming creates more climate changes, severe weather could cause the dislocation of previously stable populations. In New Orleans, one of the main causes of damage was the breaks in the levees. Levees are barriers to increases in water and are often made of reinforced earthen structures. The area of New Orleans near where they broke was a poor, African American neighborhood known as the 9th Ward. Some claim the levee broke there because the inhabitants received lower levels of municipal service provision as a result of their race. New Orleans is also a city with one of the highest proportions of African American residents, and some claim that is why they were called refugees and not victims. Victims get assistance, but refugees are on their own seeking refuge.

It is predicted that by the year 2010, there will be as many as 50 million people worldwide will be escaping the effects of environmental deterioration. This estimation is based on the 100 million people in the world that currently live at or below sea level or in places affected by storm surges. As recent events, like the tsunami in the Indian Ocean and Hurricane Katrina in the Gulf of Mexico, demonstrate, the world's coastal peoples are in danger. Furthermore, insurance companies are refusing to cover regions that are affected by these types of environmental disaster, an issue that is exacerbated by the conclusion that most of the people affected by these disaster are not in regions where insurance is even an option.

Most environmental refugees are displaced by floods, storms, and water surges, but other environmental refugees are also displaced by drought. The myriad of ecological changes affecting the global climate are also affecting the movement and ultimate displacement of settled people. Furthermore, some of these climate changes can be attributed to human-made disasters. The victims of these environmental disasters are facing tough consequences, and it is becoming increasingly important for the international community to recognize their plight.

It is in the equity aspects of sustainability that concern for the causes and consequences of environmental catastrophes lies. In areas ravaged by floods, hard decisions about relocating or staying are made in private and public areas. *See also* **Volume 2, Chapter 4: Disaster Relief and Displacement.**

THE URBAN CONTEXT OF EQUITY IN THE UNITED STATES

In the United States, colonial and postcolonial urban development first occurred around water-based transit routes. Navigable rivers provided transportation for people, goods, and raw materials. In the 1800s, railroads swept the United States, determining the location of many modern cities coast to coast. Railroads followed lowlands and rivers when they could, because it was easier to build and maintain them. Where rivers and railroads met became critical junctures in the unloading of freight and disposal of waste. With the advent of the car in the 1900s, roads were developed to become an important part of the transportation network. Like rivers and railroads, these transit decisions laid the groundwork for many U.S. cities. In this manner, over time cities became multinodal transportation networks, critical to exchanging goods and services. After World War II, Western Europe developed a more extensive network of railroads while the United States developed the Interstate Highway System. Although the interstate was built primarily for national defense, trucks moving freight used it extensively, often generating increased truck traffic around cities. Today, most freight in Europe is shipped by railroad, and in the United States most is shipped by large trucks.

These urban transit nodules developed long before any governmental environmental agencies and before most zoning law. The nodules can often be the sites for decades of accumulated wastes. Until the late 1990s, railroads in the United States dumped waste on the tracks. Ships are still allowed to empty their bilges in the port and have done so for hundreds of years. The wastes come from the transportation modalities, in terms of fuel, solvents, and animal and human waste, and from the products and raw materials themselves. As the petrochemical and chemical manufacturing industries swung into high gear in the 1940s and 1950s, the use of multimodal transportation networks dramatically increased, as did the chemical perniciousness of their wastes. This pattern of urban development is the backdrop for the application and development of sustainable programs. It generally means that industrial pollution continues to accumulate, and that the carrying capacity of the land will be reached and exceeded, and toxicity will continue its movement into larger natural systems.

These communities now face increased development and political pressure to accommodate larger waste streams from more and more places. Continued population growth of the highly consumptive U.S. population has increased waste and pressure on waste industries and cities to place waste wherever they can find space.

Of course, it is imperative that waste and pollution are cleaned up, and this is a fundamental proposition for most sustainability positions. In 1980, the U.S. EPA tried to develop indicators for locating waste and pollution, focusing first on the worst risks—controlled and uncontrolled hazardous waste sites. The EPA found it was not the soil type, nor the hydrology of the bioregion, nor the climatological conditions that determine the location of these sites. It is the race of the nearby population of humans who reside in the area that determines the probability that such a site exists. Specifically, the greater the proportion of African American people in a community, the more likely it is to be situated near such a site. For those who advocate sustainability, dumps, uncontrolled and controlled hazardous waste sites, and their transportation routes and transfer stations are fundamental barriers to sustainability. As important, searching questions are raised by sustainability advocates, the environmental burden borne by people of color in the United States becomes more and more apparent. Recognition of these burdens is not always enough for sustainability because the environment is a somber judge of results only. If recognition of the disproportionate and adverse environmental impacts on people of color is as far as sustainability policy can develop, as in housing, education, and employment areas, then it will not be sustainable because recognition of this issue is not the same as the environmental intervention and planning necessary for sustainability.

The U.S. pattern of urban development neglected environmental impacts, and they now have the accumulated wastes of over 100 years of unchecked industrialization in the world's most industrialized and consumptive nation. Environmental burdens accumulate in ways that more directly affect the self-interest of a broader spectrum of people. The public health implications of environmental degradation become harder to ignore, or hide, over time. As more and more people become more aware of pollution and its effects, they seek recourse from the government. Local government is closest to many citizens, as opposed to state or federal environmental agencies. Residents seek more control over the governmental processes that most directly affect them. Land use forums have not been environmentally receptive. Sustainability, with its equity component of inclusion, offers a more responsive forum for both the public health concerns of the citizens and for environmental planning. Unlike land use forums that can develop rules for the use of land, however, nothing yet comparable exists about sustainability that has the same force and effect as land use law.

Grappling with a legacy of industrial pollution at the local level is one of the most difficult challenges for sustainability in the United States. Land is a critical nexus because all environmental carrying capacity analyses rely on the characteristics of the land for human-involved ecosystems. The equitable component of sustainability must engage land use control because it is likely that most U.S. cities have exceeded their carrying capacity. Equity is concerned with the benefits and burdens of all past, present, and future environmental decisions. For human settlements that span working bioregions, it may be that the suburban advocates of sustainability will have to decrease the burden of an exceeded carrying capacity in nearby urban neighborhoods. The unavoidable question is: what is the carrying capacity of the land carry in terms of wastes, pollution, and development? So far, in the United States, there has not been a serious consideration of the ecosystem carrying capacity of an urban area.

References

Bullard, Robert D., and Glen S. Johnson, eds. 1997. *Just Transportation: Dismantling Race, and Class Barriers to Mobility*. Stony Creek, CT: New Society Publishers.

Davenport, John, and Julia L. Davenport. 2006. *The Ecology of Transportation: Managing Mobility for the Environment*. New York: Springer.

Goldfield, David. 2006. *Encyclopedia of American Urban History*. Thousand Oaks, CA: Sage.

Harris, Leslie M. 2002. *In the Shadow of Slavery: African Americans in New York City 1626–1863*. Chicago: Chicago University Press.

Kusmer, Kenneth L., and Joe W. Trotter. *African American Urban History since World War Two*. Chicago: Chicago University Press.

Rogers, Heather. 2006. *Gone Tomorrow: The Hidden Life of Garbage*. New York: New Press.

Urban Sprawl

It is hard for cities not to sprawl because increasing populations that consume resources in historically environmentally degrading ways tend to spread over the landscape without regard to environmental impact. Without a change in values, many cultures simply spread out and use the land, air, and water as they did when populations were small and technology used in development was limited.

Sprawl is a contemporary land use problem linked to past U.S. development patterns. A problem associated with sprawl is that people move out into natural areas with greater and greater impact on the environment and with greater consumption of nonrenewable resources. Three factors fuel sprawl. First, the tendency of people in the United States to reside away from work requires transportation back and forth to work and home. Roads and motor vehicles greatly affect the environment in nonsustainable ways. Second, our methods of commuting are inefficient, with many one-driver car trips. This also increases environmental impacts. Third, our privacy and status values are reflected in where we live, which tends to be in large-lot, low-density housing. This has a large environmental impact. In some U.S. cities there are more three-car families than one-car families. The race and income separation of U.S. residential development is a further challenge to equitable aspects of sustainability. In some communities, housing is so unaffordable that needed city workers cannot afford to live in their community. Municipal employees like fire, police, teachers, planners, and emergency response teams all do well to live in the community they serve because when the cost of living becomes too high, as in several outer ring suburbs in the United States, municipal employees, blue collar workers, and contractors must commute from suburb to suburb. This commuting pattern differs from what many transportation planners anticipated, which is the usual city center to suburb commute. The failure of U.S. mass transit planning has allowed land use to be controlled by motor vehicles. From an equity perspective, sprawl exacerbates the difference in environmental benefits and burdens between city and suburb while increasing overall environmental impact. The increase of societal concern for sustainability creates an influx of resources and brings environmental expertise to the U.S. urban land use planning process, discussed later. *See also* **Volume 3, Chapter 2: Urban Land.**

References

Benfield, F. Kaid et al. 2001. *Solving Sprawl: Models of Smart Growth in Communities across America.* Washington, DC: Island Press.

Christensen, Julia. 2008. *Big Box Reuse.* Cambridge, MA: MIT Press.

De Sousa, Christopher. 2008. *Brownfields Redevelopment and Quest for Sustainability.* Oxford, UK: Elsevier Science.

Devuyst, Dimitri et al. 2001. *How Green Is the City?: Sustainability Assessment and the Management of Urban Environments.* New York: Columbia University Press.

Dixon, Tom et al., eds. 2007. *Sustainable Brownfields Regeneration: Livable Places from Problem Spaces.* Hoboken, NJ: Wiley-Blackwell.

Helming, Katharina, Marta Perez-Soba, and Paul Tabbush. 2008. *Sustainability Impact Assessment of Land Use Changes.* New York: Springer.

Regionalization and Urban Municipalities: Patchwork Sustainability?

An important aspect of U.S. urban development is the lack of a truly regional approach. U.S. areas of human habitation are politically divided into municipalities, which derive their power from the state. There are many separate and overlapping municipalities in the United States.

Municipalities can be cities, towns, villages, counties, special districts, school districts, and any other political subdivision of the state. A large metropolitan area can have 20 to 30 municipalities. Areas that are not under the control of any municipality are called "unincorporated areas." According to 2002 data from the U.S. Census Bureau, the number of local governments was 87,525. This was an increase of 1 percent from 1997. The local governments comprise 3,136 counties, 19,429 municipalities, 16, 504 townships, 35,052 special districts, and 13,506 school districts. The average population of county government is approximately 83,000, ranging from Loving, Texas, with 67 people in 2000, to Los Angeles County with over 9.5 million people. In all, 31,877 special districts of the 35,052 special district governments served a single purpose. These special districts can overlay a municipality or county government; 21 percent of the special districts served a natural resource purpose such as fire or flood control. In terms of sustainability, these special districts may be influential. They may have to increase in number and power over ecosystem assessment and monitoring to be implemented in some aspects of sustainable development. Special districts are the closest units of government that can assess carrying capacity of ecosystems. They may be most aware of the environmental limitations and risks that application of the precautionary principle engages.

Municipalities compete with each other for a high tax base, which is economically developed from high property values. They also prefer a low cost of services. A low-rise corporate office park would be preferable to a low-income trailer park with many children to most U.S. municipalities that could exercise a choice. Most economic development requires that business and industry create a commercial presence, and industry generally shies away from strict local environmental controls because of fear of loss of profits. The competition between municipalities to increase their tax bases and economically develop without environmental controls creates large differentials between them in the areas of education, environment, transportation, housing, and employment opportunities. This is sometimes called a "race to the bottom" and is usually used to describe states that lower environmental requirements to attract industries. Research comparing states with low environmental requirements with states with strict environmental regulation demonstrates that states with strict environmental regulations had a higher quality of life, including economic development, then states with the lowest environmental enforcement. This is an awkward landscape for the inclusionary aspect of equity in sustainability to apply to a given bioregion. One recent development in U.S. land use, however, may be a bridge to sustainability. *See also* **Volume 3, Chapter 3: United States.**

Smart Growth: Smart Enough for Sustainability?

Ideas such as "smart growth" are now being applied as a way to combat sprawl. Smart growth seeks to limit development to a balanced approach that incorporates issues like affordable housing and environment; however, it is being slowly considered only on a municipal basis. If urban areas have exceeded their carrying capacity, smart growth would require suburban areas to balance out some of the difference in total environmental load by accepting more industrial development, waste sites, waste transfer stations, and other locally unwanted land uses. It would probably require a large amount of environmental cleanup and environmental land and water reclamation for most U.S. cities. Another practical limitation on smart growth approaches is the law. If a regulation takes away the complete market value of land, it is a taking of private property under the U.S. Constitution. If it is deemed a taking by a court, the government must pay the landowner the fair market value of the land. Generally, local governments do not like to have to pay via takings if they can avoid it because it would make projects too expensive.

Reference

Porter, Douglas. 2002. *Making Smart Growth Work.* Washington, DC: Urban Land Institute.

Environmental Costs of Urban Exclusion

In the United States, the land with the most intense human footprint, the land that has the weakest enforcement of environmental laws, and the land that will be necessary for sustainability is decidedly urban. In addition, in the United States this is one of the most ignored aspects of public and private environmentalism. Most of the pollution is located in cities. In the United States and in many cities around the world, cities are points of immigration and migration. There are high populations of people of color in the United States. The federal government began a strong program of environmental regulation around 1970 when the U.S. EPA was formed. Cities were not prioritized, as they were in many U.K. countries.

U.S. environmental policy does not require environmental impact assessments for most activities in cities for several reasons. Sometimes the U.S. EPA or other federal agency has determined that cities have no complete ecosystems. This means, according to them, that if there were not a complete ecosystem, then there could be no significant environmental impacts. According to the National Environmental Policy Act of 1970, if there are no significant impacts on the environment, no full blown environmental impact analyses need be done. Another reason for the exclusion of U.S. cities from environmental protection is more political. Congress, and sometimes federal agencies, has carved out "categorical exclusions" from environmental impact programs. For example, a primary source of revenue flow from the federal government to cities comes from the Community Development Block Grant Programs. These funds can be used for many projects, some of which have significant environmental impacts. This grant program was categorically excluded from environmental impact requirements by law. Over the years many exceptions to environmental impact statement requirements have developed. Environmental impact statements are advisory only (see Volume 1). This has crippled the ability of developing an environmental baseline for sustainability in U.S. cities. Some states, such as California and Washington, have versions of environmental impact statements. These, too, have many exceptions and are advisory only. On the rare occasion federal environmental impact statements are litigated, they can take a racial turn. In one case in Chicago, African American people moving into neighborhoods were considered by the residents to have a significant environmental impact.

In 1972, the Chicago Housing Authority proposed scattered sites for its public housing. The residents of public housing were often African American and the sites were in white communities. A group called Nucleus of Chicago Homeowners Association sued the Chico Housing Authority to prevent them from locating the public housing in their neighborhoods because it violated the requirement under the National Environmental Policy Act that an environmental impact assessment be performed. They alleged that the public housing tenants had more crime, disregarded private property, and had a lower commitment for hard work. This would have a significant and adverse impact on the aesthetic and economic environment. This case went into a six-week trial with expert witnesses and much media attention. The judge eventually ruled that environments are pollution; people cannot be pollution, and they cannot be considered pollution because of their race. The National Environmental Policy Act did not include people as pollution.

References

Bartlett, Peggy, ed. 2005. *Urban Place: Reconnecting with the Natural World.* Cambridge, MA: MIT Press.

Gottlieb, Robert. 2007. *Reinventing Los Angeles: Nature and Community in the Global City.* Cambridge, MA: MIT Press.

Kible, Paul Stanton. 2007. *Rivertown: Rethinking Urban Rivers.* Cambridge, MA: MIT Press.

Equity Impact Analysis

There are proposals to develop an Equity Impact Analysis similar to an Environmental Impact Analysis (EIS). These tend to follow the same basic structure as an EIS but with a focus as a tool for local policymaking. A common ground for measuring adverse effects here is federal Executive Order 12898. Adverse effects can be controversial from an environmental perspective. From an equity perspective, they mean the totality of significant individual or cumulative human health or environmental effects, including:

- Interrelated social and economic effects, which may include, but are not limited to bodily impairment, infirmity, illness, or death

- Air, noise, and water pollution and soil contamination

- Destruction or disruption of man-made or natural resources

- Destruction or diminution of aesthetic values

- Destruction or disruption of community cohesion or a community's economic vitality

- Destruction or disruption of the availability of public and private facilities and services

- Gentrification or displacement of persons, businesses, farms, or nonprofit organizations

- Increased traffic congestion, isolation, exclusion, or separation of minority or low-income individuals within a given community or from the broader community

- Denial of, reduction in, or significant delay in the receipt of, benefits of programs, policies, or activities

Equity impact assessments tend to be focused on local actions such as land use. In terms of an equity analysis, the question of where degrading or risky land uses are allowed and prohibited can affect the adverse impacts the residents experience. Proposed policies with adverse impact potential would be subject to analysis that asks how the policy affects access to jobs, housing, health care, and education. Would the proposed action degrade the quality of life? What would the burdens and benefits be from the policy for each citizen or neighborhood?

Like an environmental impact assessment, an equity impact assessment would define its scope, involve stakeholders, consider draft alternatives, and seek to mitigate impacts. In terms of the equity component of sustainable development, equity impact assessments would be useful because they increase participation, focus on actual land uses, and increase knowledge of ecological impacts.

The cost of excluding cities and their communities from environmental and land equity is a depletion of natural resources. Clean air and clean water are natural resources necessary for a healthy life and they affect regions near and far. Another cost to the equity equation for many city residents and sustainability advocates is that of growing environmental degradation that cannot be escaped by the invocation of privilege or money. Many sustainability advocates fear that current U.S. expensive and time-consuming litigation of antiquated environmental laws creates unnecessary adversaries out of neighbors. This can create a lack of inclusion and fairness when environmental and human concerns are not enough to create a sufficient case of action for judicial intervention.

One primary characteristic of an equity issue is that the consequences of environmental decisions are inescapable for those affected. Although this is a dangerous consequence for socially marginalized people in environmentally marginalized land, with the rise of sustainability it is now dawning on the powerful and privileged parts of U.S. and global society that some environmental privileges are just that—privileges—and they have come at a cost to others. Some argue this is an age-old dynamic of human habitation because someone has to live downstream, meaning there has to be some environmental degradation somewhere. With rising populations, decreasing natural resources, and accumulating pollution pockets, the consequences of environmental decisions are becoming less escapable. Sustainability is the concept that pushes realization of this dynamic for many stakeholders. In some ways, the environment itself is an uncompromising reflection of all our actions.

The Environmental Contributions of Cities

Cities can reduce human impacts on the environment by increasing the density of human habitation and using systems to control impacts, such as sewers and land use control. If the same urban population were spread out over the landscape, the impact on the environment would be greater because of the impacts of increased transportation and less efficient waste treatment.

High densities in residential, commercial, and industrial patterns of living often have lower costs per person or place for water, waste disposal, fire, police, emergency response services, education, communication, and health care systems. Another environmental advantage of urban areas is that the sources of consumption of many goods and services are near their sources of production. In theory, this decreases the amount of transportation and increases the efficiency of the transaction.

Another environmental advantage commonly attributed to urban areas is the decreased use of land. Urban areas take very little land mass to support dense populations. Although their ecological footprint, the amount of resources consumed, may be large, their actual land mass requirements are small. Cities can extend far into the sky or deep into the ground. Many futuristic designs for self-contained cities have them soaring high in the air.

Sky City in Tokyo

Globally, human populations are growing and becoming more urbanized. In Japan, this is happening at a rapid pace. Tokyo is the largest city in Japan, with over 12 million people; in fact, 10 percent of the population of Japan lives in Tokyo. It has some of the highest population densities in the world. It is also a wealthy country with a reputation for advanced technological innovation. The lack of land to accommodate growth in population, combined with the ability to conceive new solutions, led to the development of plans for a "Sky City."

Sky city is gigantic in scale. It is three tall towers connected in a triangle. They will be about 1,000 meters tall, taller than any building in the world. They will be built to handle some of the extreme aspects of the environment there, such as high winds, by using rounded shapes and having gaps so that the wind could blow through the towers. To handle earthquakes, Sky

City would be designed in a rough pyramid structure. To handle the sway from the winds and earthquakes, designers plan to put dampers in the structure to absorb the force of the wind and ground shocks. It will be designed as a self-contained city, housing 35,000 residents and 100,000 workers. It would grow its own food and have its own municipal services such as fire, police, and education.

By developing urban habitats that grow up and not out, many current environmental impacts of urban sprawl would be solved. These projects may not always be affordable in developing nations, so whether they would fully engage some of the poverty issues in the equitable component of sustainability is still in question.

Reference

Extreme engineering, Discovery Channel, dsc.dis covery.com/convergence/engineering/sky city/interactive/interactive.html.

Reference

Jacobs, Jane. 2000. *The Nature of Economies.* New York: Vintage Books.

U.S. Urban Conditions

Citizens living in urban, poor, and people-of-color communities in the United States are currently facing a wave of gentrification and displacement, as well as increased exposure to toxic substances and waste.

As part of this process, municipalities, urban planners, and developers are accomplishing much of this largely beneficial "revitalization" by taking advantage of federal policies and programs. Displaced residents and small businesses sometimes complain that revitalization is being built on the back of the very citizens who suffered in-place, through the times of abandonment and disinvestment. Downtown renovation is a matter of public policy in most U.S. cities. *See also* **Volume 3, Chapter 2: Urban Land.**

Reference

Angotti, Tom. 2008. *New York for Sale: Community Planning Confronts Global Real Estate.* Cambridge, MA: MIT Press.

How Is Urban Gentrification Inequitable?

Gentrification takes place in an urban landscape of rental communities that experience transiency. In the United States, one can either own or rent property. Most poor people and many middle-income people rent apartments until they can afford a home. In many cities in the world, many residents simply occupy the space, sometimes called "squatting." Squatting is illegal but tolerated. In some countries, such as Costa Rica, a landowner is responsible for the squatters; however, environmental and public health conditions can be deplorable. Renters have much more security in their space, but far less than private property ownership. Poor and discriminated people are often forced to move from rental property to rental property and experience homelessness and overcrowding at a much greater rate than the general population. Now that these areas are being cleaned up, whether for sustainability or increasing market demand for clean urban property, property ownership is displacing renters.

Gentrification is generally thought of as private market, piecemeal renovation. As the waves of new "gentry" move to large-scale renovation projects in or near central business or warehouse districts, they come into direct competition with the current residents of these formerly forgotten places. Many of these older urban areas suffered from the industrialized waste practices of the past and were not in high demand for residences. Low-income people, recent immigrants, and people of color who were unable to find or afford shelter elsewhere set up communities in these areas. The commodity of land being sold in the real estate market is more than a physical structure or piece of acreage. It is also a neighborhood—a political and cultural entity necessary for the sustainability of a community in that place. Gentrification has placed poor populations into direct competition with relatively powerful and privileged populations in urban areas for inner city space. Environmental cleanup of these formerly industrialized, now residential, communities can be a powerfully displacing force in this context. Without knowledge of the past and present land use practices, neighborhood culture, and localized financial lending patterns, the EPA may unintentionally exacerbate the displacement of low-income and people-of-color communities by its emerging sustainability policies and practices. Being inclusionary will help federal environmental agencies avoid displacing environmental policies and thereby environmentally empower communities to be part of long-term ecological solutions.

The environmental cleanup of urban areas is a high priority for any serious policy of sustainability. In the United States, cleanup policies are still in their infancy, characterized by timid environmental enforcement, weak environmental mitigation, incomplete ecological and cumulative risk assessment, and many polluted sites. Although cleanup is a high priority, it may be a difficult challenge because of the history of environmental neglect. As noted in the early 1970s:

> People in cities bear the brunt of technological and urban sprawl—in pollution and resulting disease, auto–dominated transportation, inadequate housing, and dangerous, degraded neighborhoods.

National Environmental Justice Advisory Council. 2006.
Unintended Impacts of Redevelopment and Revitalization Efforts

in Five Environmental Justice Communities, p. 3 (online at epa.
gov/compliance/resources/publications/ej/nejac/redev-revital-
recomm-9–27–06.pdf)

The EPA, however, did not vigorously engage urban areas until cleanup
became mandated by law in the 1980s. Wastes had accumulated in
urban areas to such an extent that they could no longer be ignored. In
2004, the U.S. Office of the Inspector General recommended that the
EPA examine unintended environmental and community impacts like
gentrification. They concluded:

1. The urban environmental problems identified by the EPA,
   community groups, and environmental organizations in the
   1970s *never went away.*

2. Their impacts and risks began to accumulate noticeably.

3. The accountability of state and federal agencies for environmental justice increased.

The beginning programs for environmental protection of the 1970s re-
mained as they were before urban interventions in the early 1990s. The
new programs for sustainability will address a range of pressing urban
environmental issues to be relevant in all cities, including U.S. cities.
First, city residents want to reduce chemical and physical environmental
hazards where they reside, work, play, worship, and travel. In most major
metropolitan world cities, a primary concern is water supply and quality, especially as it relates to public health and infectious diseases. Waste
treatment is intrinsically related to water quality and quantity. Overdevelopment in limited aquifers reduces water quantity and quality for
long-time, current, and future residents. With rapidly increasing global
and urban population, water issues loom large.

Gentrification, or moving human populations around in urban
areas, does not necessarily decrease environmental impacts or address
poverty issues. By making better environmental conditions available to
those who can afford them, gentrification moves poor people into environmental conditions that can pose a threat to both public health and
ecosystem preservation.

References

Aitkinson, R. *Gentrification in a Global Context: The New Urban Colonialism.* Andover,
    UK: Routledge.
Lees, Loretta et al. *Gentrification.* Andover, UK: Routledge.
Sze, Julie. 2006. *Noxious New York: The Racial Politics of Urban Health and Environmental
    Justice.* Cambridge, MA: MIT Press.

Ecosystems and Urban Populations

Other environmental issues likely to dominate early inclusionary dis-
courses about sustainability in urban areas relate to softening the
ecological footprint of cities. These approaches want to minimize the

transfer of any environmental impacts on people and ecosystems of the city. Ecosystems of a city can, and usually do, cross political boundaries of nations, states, cities, and any other human-derived arrangement of geography. An ecosystem is a community of animals and plants interacting with each other and the physical environment. They include physical and chemical components such as soil, water, and nutrients of organisms there. Humans are major parts of ecosystems, with major impacts, and with a new concern for sustainability. The list of human impacts on the environment is long and will be revisited many times as modern ecosystems are reconstructed. Human impact on the environment is conceptualized as the interaction between humans and their ecosystems. Ecosystems are the basic building block of human and environmental interaction, and are applied to cities.

An ecosystem is a dynamic group of animal, plant, and microorganism structures in their nonliving environment. These structures interact as a functional unit. Ecosystems have many complex interactions between their various life communities, but they have vague and moving boundaries. The scientific concept of ecosystem is relatively new. Arthur Tansley and Raymond Lindeman did the first quantitative ecosystem study in the early 1940s. The first textbook around the ecosystem concept was published by Eugene Odum in 1953. It is a relatively new concept in science and an even newer concept in the management of natural resources. Its changing boundaries and newness make it difficult to study in an urban context. Human impacts on ecosystems are fundamental to understanding sustainability, but they also make it difficult to know the size, health, and boundaries of modern ecosystems.

Humans view ecosystems in terms of their ecosystem services, as opposed to the intrinsic value of the ecosystem itself. This is particularly true in an urban context. Ecosystem services can be divided into three main groups.

The first group can be labeled as *provisioning*. This refers to the provision of food, energy, and fresh water.

The second group is referred to as the *regulating* function of ecosystems. Ecosystems strive to maintain a balance among its components, but cities often throw this balance into disorder. The regulating ecosystem service refers to air and water purification for life, mitigating floods and droughts, maintaining soil fertility, enriching the range of biodiversity, and stabilizing changes in climate. With global warming and climate change, the regulating functions of ecosystems are taxed. Add to that concern the increasing and urbanizing populations of poverty and the threat to systems of nature on which future life depends can increase. This in turn increases the potential application of the precautionary principle to additional development proposals, which can decrease the potential for economic development in areas of poverty.

The third ecosystem service of ecosystems is that of *meeting human social needs*. Nature provides many intangible services to human needs that include beauty, spiritual reflection, education, and scientific research.

All cities have these needs and when they are ignored, the quality of life can decrease. This can have an effect on overall public health, as well as on attitudes toward the environment that can hamper ecosystem preservation and protection.

Population Undercounts in Cities: Why Does It Matter for Equity and Sustainability?

If environmental policymakers are unaware of human consequences of the application of environmental policy to urban areas, then displacement and gentrification of neighborhoods can occur. The usual unintentional way to avoid communities is by population undercounting, dilution of salient race demographics, or disappearance (as a result of undercounting) of people from official population counts. Then it is easy to claim that consequences to these populations were not intended. In the inclusionary dialogues that characterize equitable aspects of sustainability, these unintended consequences will be a high priority for many urban communities. Many U.S. cities have contested the census for undercounting minority populations, primarily in urban areas and most often with people of color. The population estimates of the world's megacities often omit squatting communities of poor people. These "invisible" people are affected by degraded environments and often view current environmental processes as unfair. When some models of urban sustainability stress more inclusion, equity and public health concerns can rise to the fore because of these and other views from previously excluded people. In many communities of color, however, population counts by the government may be avoided. Concern about arrest, deportation, or harassment from government authorities is a real obstacle for the inclusionary dialogues that surround some versions of sustainability.

The majority of urban dwellers are people of color. Together with people of color, women, and children will compose the majority of the world's population, increasingly in cities. As the deterioration of our ecosystems accelerates, these people suffer most (see Millennium Ecosystem Assessment (MA), Environment and Ecology, Appendix B.). Many climate change models predict the greatest impacts on the equator, where growing populations already face periodic food shortages. Equatorial ecosystem dynamics have global implications in terms of climate change and ocean currents. This area is urbanizing at rates never before known in human history. As urban human habitations become more central to global sustainability efforts, their histories of human oppression become known. Inclusionary dialogues form an important part of the equity component of sustainability. Advocates of sustainability will have dialogues around past oppressions, as environmental histories are unraveled to create an environmental foundation for sustainability. This inclusion may spread to other traditional forums like land use or municipal service provision. As cities programmatically develop beyond traditional sanitation and garbage pickup to recycling and eventually part of sustainability programs, these formerly excluded areas will demand remedial environmental restoration, at least to levels of safe public health. The issues of racism and cultural exclusion can become accepted parts of major social institutions. They can seem accepted and therefore "normal," making issues of oppression invisible. Daily life considerations affecting waste and what is produced as waste may now be invisible to some groups and not others. These dynamics present formidable obstacles in creating an inclusionary type of environmental decision-making structure and process necessary for meaningful policy development of sustainability. The overall context of equity and sustainability is the land, air, and water. Land in urban areas in some industrialized nations, like the United States, has not been included in modern environment developments and advocacy. Despite the obstacles, the increase in overall public awareness of human health conditions in urban areas has led to an increase in residents' demands for notice and involvement with major land use and environmental decisions and accountability. One of the best sources of past and present environmental conditions and land uses is the residents of the community. The need for real-life, accurate environmental baseline data is greatly facilitated when the community has the capacity, training, and access to resources to assist.

Reference

Principles of Environmental Justice. www.ejnet.org/ej/principles.html.

Millennium Ecosystem Assessment

The Millennium Ecosystem Assessment (MA) is, to date, the most comprehensive survey of the ecological state of the planet. The MA was called for by then–Secretary General of the United Nations Kofi Annan in 2000 and began its work in 2001. The MA was undertaken by an international network of scientists and other experts. More than 1,300 authors from 95 countries were involved in the MA and were divided into four working groups. Three of the working groups, Condition and Trends, Scenarios, and Responses, focused on global assessment goals whereas the fourth working group focused on subglobal assessments. The resulting assessment was divided into four technical volumes that were reviewed by experts and governments, and more than 600 individual reviewers worldwide provided approximately 18,000 individual comments. The assessment lasted four years and concluded in 2005 with its report on the consequences of ecosystem change for human well-being and the scientific basis for actions needed to enhance the conservation and sustainable uses of the Earth's resources. The MA reported four main findings:

1. Over the past 50 years, humans have changed ecosystems more rapidly and extensively than in any comparable period in human history, largely to meet rapidly growing demands for food, fresh water, timber, fiber, and fuel. This has resulted in a

substantial and largely irreversible loss in the diversity of life on Earth.

2. The changes that have been made to ecosystems have contributed to substantial net gains in human well-being and economic development, but these gains have been achieved at growing costs in the form of the degradation of many ecosystem services, increased risks of nonlinear changes, and the exacerbation of poverty for some groups of people. These problems, unless addressed, will substantially diminish the benefits that future generations obtain from ecosystems.

3. The degradation of ecosystem services could grow significantly worse during the first half of this century and is a barrier to achieving the MDGs.

4. The challenge of reversing the degradation of ecosystem while meeting increasing demands for services can be partially met under some scenarios considered by the MA, but will involve significant changes in policies, institutions, and practices that are not currently underway. Many options exist to conserve or enhance specific ecosystem services in ways that reduce negative tradeoffs or that provide positive synergies with other ecosystem services.

The findings outlined the negative impact of human actions on the Earth's natural capital. The MA showed that humans cannot take the Earth's ability to sustain future generations for granted at this rate of environmental degradation; however, if appropriate substantial action were to be taken it would be possible to reverse some degradation over the next 50 years. The MA is an assessment of data that was already available; it is unique in that it was a global assessment that presented a consensus view of the current state of the planet. Consensus is one of the cornerstones of change. Furthermore, the MA also identified a number of "emergent" findings or conclusions that were the result of examining a large amount of information together.

One of the simultaneous strengths and weaknesses of the MA is the gaps in knowledge about the status of the planet's ecosystem. There is relatively limited information available at a local and national level about the status of ecosystem services and even less information about the value of nonmarketable services. There is also limited information about the economic costs of ecosystem degradation. In an increasingly market-based world, it is imperative for economies to have information and assessment of economic loss of human action. The MA created awareness of these gaps in knowledge and stimulated data gathering and assessment to eliminate the gaps.

The MA work groups expect that there will be significant adoption of the MA conceptual framework that will continue to help meet the planet's sustainability needs and reverse degradation. The MA indicated

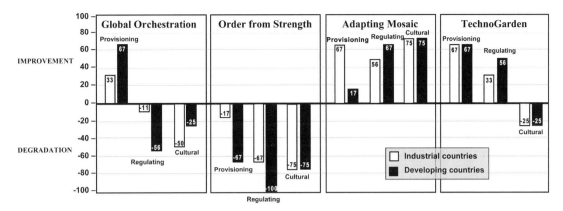

FIGURE 3.4 • Number of ecosystem services enhanced or degraded by 2050. A 100 percent degradation means that all the services in the category were degraded in 2050 compared with 2000, whereas 50 percent improvement could mean that three of six services were enhanced and the rest were unchanged or that four of six were enhanced and one was degraded. The total number of services evaluated for each category was six provisioning services, nine regulating services, and five cultural services. Philippe Rekacewicz, Emmanuelle Bournay, UNEP/GRID-Arendal. http://maps.grida.no/go/graphic/number-of-ecosystem-services-enhanced-or-degraded-by-2050.

that changes be instituted firmly and quickly. It was recognized that, as humanity has the power and ability to prevent the damages to the planet, it also has the duty to do so. One of the most important issues brought up was the effects of environmental damage to the underdeveloped and poor people of the world. The report urged the nations of the world to work harder to achieve a sustainable future.

WHAT VALUES ARE SUSTAINABLE?

Values have an economic and moral meaning, both of which are reflected in how a society organizes its economic concerns and distributes its resources. Definitive values tie together the entire concept of sustainability. First, there is an explicit concern for future generations. The major documents on sustainability all refer to a principle of resource use that protects the interests of future generations of humans and nonhuman species. The central defining characteristic of sustainability is its concern that contemporary humans conduct themselves in a way that protects the interests of future generations of humans: a specific kind of intergenerational equity. In this way, intergenerational equity is the central, core value of sustainability. This is reflected in a variety of ways including the concept of carrying capacity, sustainable development, and resource management.

Values of Industry

By contrast, the values of industrialization have added to the complex nature of the environmental and equity problems that challenge the development of sustainable programs and policies. This is especially true when considering the political and social decisions during the age of industrial development and European urban settlement. These values include the use and potential exploitation of natural resources in pursuit of profit, with profit deemed a proxy for a social good. Profit as a motive is a powerful force and has arguably provided a level of goods and services that has extended the quality and length of human life. An economy based on the idea that profit is a proxy for an unquestioned social good can become an economy operating in contradiction to sustainable approaches and practices. For example, critics point out how these values ignore the creation of waste and pollution, and their inevitable environmental consequences on the planet's ecosystem. Examples include the dumping of treated sewage into rivers and oceans, the creation of radioactive waste as a by-product of nuclear energy generation, and the release of the greenhouse gas carbon dioxide along with the carcinogen dioxin as a result of the incineration of certain petrochemical products.

In addition, the economies of most industrialized nations, including the United States, are driven by consumerism that is deemed necessary to drive an economy based on maximum use of natural resources for as much profit as possible. The policies, customs, and personal practices founded on industrial values have led to a rapid diminution of natural resources, geometric increases in pollution problems, and near gridlock in the ability of contemporary policymakers to address these policies-driven, value-laden, postindustrial environmental problems. With rapidly rising human populations and more pressure to consume natural resources for economic development, the overall damage to the planet's ecosystem could have devastating consequences. This is especially a concern for marginalized populations.

The values that lay the foundation for sustainability are of explicit concern for future generations of life in all its diversity and questioning consumerism. The values of sustainability are underscored by the realization that it is no longer possible to relentlessly and recklessly exploit natural resources on which all life depends. Industrial values and the policies they formed led to our current postindustrial environmental gridlock, especially in matters regarding sustainability and equity. Because many Western industrialized societies are in decisional gridlock on environmental issues, and the environmental degradation threatens life on which we all depend, more people become aware of the values that underlie industrialization and uncontrolled profit-taking from natural resources. When the environment becomes "personal," more people value sustainability. Many communities of color in the United States often describe their involvement with the environment as personal, although it is often because of the health risks it imposes on them.

Reference

Roodman, David Malin. 1996. *Paying the Piper: Subsidies, Politics, and the Environment.* Washington, DC: Worldwatch Institute.

Private Property: A Strong Value Tied to Liberty

Ownership of property gives individual owners the power to determine what choices to make regarding that property in some cultures. This right may exist without knowledge of, or regard for, the consequences of that individual choice for the ecology of the place. To the extent that government ensures such absolute expectations of rights to property, sustainable choices for environment and for community may be frustrated.

One of the strongest values in the United States and many other nations is the ability of a citizen to own land. In the United States, private property is equated with life, liberty, and the pursuit of happiness as stated in the U.S. Constitution. Actually, private property is historically and comparatively a much more flexible concept than a rigid and inflexible right. Land, or real property, is a bundle of rights, responsibilities, and privacy expectations. U.S. ideas of land ownership came via the colonialists from England. In the 1600s and 1700s, private property owners were a small privileged class. All judges and voters owned property. Most jurists in jury trials were property owners. Large landowners exerted substantial political power in the postcolonial development of the United States. To Western European settlers, land seemed limitless—there just for the taking. Land they perceived as manifest destiny to own and as wilderness, however, actually was already occupied by indigenous people.

In the United States, 3 percent of the land is owned by industries and 2 percent is owned by private residences. Overall, there are about 2.1 billion acres, which is about 7.7 acres per person, 14 times higher than the rest of the world and three times higher than any other industrialized nation. About 16 percent of that land is occupied by 341 Metropolitan Statistical Areas. The federal government is a large landowner, owning 96 percent of Alaska, 86 percent of Nevada, 66 percent of Utah, 63 percent of Idaho, and 52 percent of Oregon, as well as substantial acreage in other states. This represents about 730 million acres, or about one-third of the total land and water area of the United States. The ability of the state to control the use of land, whether public or private, leased, owned, or held in trust, is important for any policy of sustainability.

Where did this strongly held value and imported Western European concept of "private property" come from? Private property owners may not willingly give up property rights, even if it is for sustainability. The inclusionary aspect of the equity component would include private property owners. Residential, commercial, and industrial owners of real property all have a strong stake in this process. The settled expectation of what rights make up "private property" is well protected by state and federal constitutions. Understanding the roots of this value is important for sustainability because the settled expectation of private property owners may be forced to change.

For many of the first U.S. colonialists, the concept of private property came from John Locke. Locke made "natural law" arguments for the creation of private property. He challenged monarchial absolute sovereignty with the idea that God gave the Earth to humans in general, and not to specific individuals. As the basis for the legitimacy of their rule, monarchs (kings, queens) claimed lineage with specific recipients of God's glory. By challenging monarchial claims to absolute, God-given sovereignty over all land, Locke gave impetus to the idea that those others than the king could own land. Individuals not associated with the government could have land with the same set of expectations as government land. Locke wrote that God, by making humans in his image, gave value to nature. Nature, the environment, had value to the extent human's labor gave it value. In his view, everything existed for humans as part of God's grand plan. The settler's standard for being human, however, did not incorporate many people of color. "Savages" and slaves were not considered human and therefore were unable to own property, but they were able to *be* property. In early U.S. history, only white men over the age of 21 could own land and vote, and sometimes they had to own land just to register to vote. As industrialization and capitalism grew with the fledgling U. S. democracy, it became clear that the laborer did not own the land on which she or he labored.

Locke's theory starkly failed to include everyone when applied by the colonialists because he underestimated the depth of inhumanity, the rapid rate of industrial and technological development, and the limitation of the environment. In terms of Lochean land theory in application to the United States, African slavery, the ecocide of indigenous peoples, and the modern exploitation of Latino, indigenous, and African American farm workers refute the old idea that those who labor on the land create value they own. With rapid frontier expansion and the ecocide of indigenous people, it was easy to overlook the consequences of this action on both land and people, and the industrial value of private property grew strong and thrived. Land value became considered only in terms of economic value of it or its products.

The theoretical marketplace is not the whole community. A community may die while a market thrives, especially in a modern era of global markets that include land. Multinational corporations seeking cheap natural resources may irrevocably decimate the ecosystem and public health of a given community, but make enough profits to justify these actions to noncommunity investors, shareholders, land speculators, and governments. Land has many other values that are not with a price. Some industries are particularly concerned with sustainability proposals because they may be currently unable to act sustainably. It may be difficult to mine for gold or oil in a fragile desert ecosystem in a sustainable manner, for example. Some deep ecologists feel that to even put a price on land devalues it, and that humans are caretakers of the land for the environment and for future generations. Other values tied to land can include spiritual, cultural, historical, environmental, and sacred. The older values of preindustrial communities are often the new values of sustainability. Places are valued for their human and community habitability, for the strength of the ecosystem, for their diversity and resilience, which are not necessarily reflected in market value.

Over the years, the bundle of rights associated with private property has shifted. Land use law and planning, discussed later, prevent pure market forces from determining all uses of land. In the United States and in most places in the world, environmental law is relatively new and poorly enforced. Most of the time industries are allowed to self-report emissions, citizen evidence is not allowed to challenge a given environmental permit, and urban areas are ignored. Environmental law and regulation are barely a challenge for private property regimes. Sustainability may present a serious challenge to private property regimes and require that the bundle of rights again be altered. It is a challenge that gets larger every day.

Private property is not a rigid concept, but a fluid concept of law that changes over time based on the needs of the community. It is re-defined, defended, altered, and confirmed in the courts. Most of the ideological and political activities of U.S. courts in land use are focused on adjusting the boundaries between public and private interests. The result of these conflicts are the laws and rules that both detract from some of the rights and privileges of private property owners, and secure them from incursion from the government and others. Private property rights in the United States are continuously refined and confirmed in courts, legislatures, government agencies, business transactions, and in-heritances. The contours of the right and the value of private property change to reflect the needs of local, regional, and national communities. Technological advances in our ability to affect the environment have dramatically increased the chances of committing irreversible errors that may limit rights of both future generations and current expectations of private property owners. Advances in ecosystem knowledge, increases in the technology of environmental monitoring and risk assessment, and new policies of sustainability may require a shift in the bundle of human rights in land called private property.

References

Freygole, Eric T. 1993. *Justice and the Earth: Images for Our Planetary Survival.* New York: The Free Press.

Freyfogle, Eric T. 2003. *The Land We Share: Private Property and the Common Good.* Washington, DC: Island Press/Shearwater Books.

Raymond, Leigh Stafford. 2003. *Private Rights in Public Resources: Equity and Property Allocation in Market-Based Environmental Policy.* Washington, DC: Resources for the Future.

New Values

Economic and efficiency values of industry and the free market have been considered value neutral and have always been a part of the decision-making process in industrialized nations. Their true value neutrality is now challenged as simply convenient ignorance. Because of these values, some sustainability advocates argue, environmental protection policies have long excluded people, their cities, and the centuries of accumulated pollution from much serious attention.

Sustainability and environmental equity require that ecosystems be dealt with as an undivided whole, with no part being unsustainable. Sustainability and social policy also require that the population be treated as an undivided whole. To treat any one point of a bioregion as acceptably unsustainable, or to treat any group of people as acceptably unsustainable, contradicts the values in sustainability because of the fundamental interconnectedness of bioregions and communities.

The study of ecology has grown tremendously and our knowledge base as applied to the quality of the environment now includes ecosystems and cumulative impacts. Ecology can give a path to knowledge to become sustainable, but not the political will. The interdependencies of complex bioregions and ecosystems are numerous and varied.

The natural order of any given place quickly overwhelms scientists mired in statistical models of causality and species specialization. The processes of sustainability have decision-making functions that fill in knowledge gaps produced by theoretical and scientific development of ecological theories with actual information from real places. The more people know about accumulating chemical emissions, discharges, fugitive emissions, bioaccumulations, cumulative risk, unregulated industries, unenforced environmental laws, and environmental impacts on the ecology and public health of people, the more becomes known about the environmental devastation of urban spaces. Hard questions about past and present actions are unavoidable and painful. A strong value of sustainability, however, is that consideration of the sustainability of given environment cannot occur without considering the sustainability of the community.

The term *sustainability* implies many things to many people. It has been used and conceived as a term by different people over time to achieve or promise a variety of results. This also prompts controversial concerns that the term is used to "green wash" practices and programs labeled "sustainable" but without real content. Different iterations of the term *sustainable development* can become vague and abstract when removed from tangible environmental and social indicators. Generally, sustainable development is a call for constraints on unlimited economic growth. It implies human duties to future generations of human life. Sustainable development is generally seen as requiring industry and commerce to internalize its waste and pollution and pay the true costs for resource use. A sustainable social policy in the United States will require major changes in consumption and production patterns of goods and services, changes in values about wealth and prosperity, a balance of moral concerns and economic efficiency, and the fair distribution of environmental risks between current and future generations, as well as between contemporary generations. Given the broad challenges of sustainable development to the status quo and to the traditional and customary way of doing business and approaching nature, old and new questions of fairness and equity arise. Some of these questions are controversial and confrontational. Ultimately, many questions about equity are dependent on values. These values can be expressly or implicitly held and can be expressed in law, policy, and all social institutions.

References

Capra, Fritjof. 1996. *The Web of Life: A New Scientific Understanding of Living Systems.* New York: Anchor Books.

Maser, Chris. 1999. *Ecological Diversity in Sustainable Development: The Vital and Forgotten Dimension.* Boca Raton, FL: Lewis Publishers.

CHAPTER 2

Definitions and Contexts

THE LANGUAGE OF SUSTAINABLE COMMUNITIES

The use of the term *sustainable* as encompassing a way of adapting human life to ecosystem limits is relatively recent. It is connected to the environmental consciousness movement of the 1960s and to a series of United Nations conferences and reports beginning in 1972 in Stockholm Sweden, and culminating with the Brundtland Report in 1987 (see Sustainable Development in this chapter). The idea has been broadly adapted to many uses that reflect competing and sometimes conflicting value choices. These multiple idiomatic uses of the language of sustainability make the idea seem ambiguous. To some extent, that may be an important political value of the language of sustainability: its appeal across difficult value conflicts. But it may also be the weakness of the many usages of this terminology.

In an effort to reveal the basis for different usages of the term *sustainability* or *sustainable development*, we have identified significant contexts for the use of this language in the three main domains that are joined by this term: Environment and Ecology, Business and Economics, and Equity and Fairness. The context for definitions provides a broad picture of how the term is used in each area. By placing these usages within a specific context, one can better understand what value choices are being posed and how they are resolved.

Also, because each domain itself has evolved a specialized and exclusive set of references to sustainability, terms and their usages are rapidly becoming less accessible to a nonspecialized public audience who is nevertheless interested in knowing about them. Sustainability is a term that can change over time as different languages and cultures identify its meanings, but generally the meanings of sustainability can be translated into ecological, economic, and equitable terms. Some have argued that linguistic diversity and biodiversity are highly correlated, and perhaps some causal relationship between the two may exist. Linguistic diversity is the ability to use different languages and terms to describe and explain an event or action. Biodiversity is the diversity between and among species and the diversity within and between ecosystems. Sustainability may unfold in language in ecosystems that are diverse.

CIVIL SOCIETY MOVEMENT

Civil society as a movement recognizes the roles of nonprofit organizations. There can be many different types of nonprofit organizations, and they are given special tax treatment in the United States. They do not necessarily have to be environmental organizations, but many of those organizations take a lead role in advocating for sustainability. Civil societies around the world are those organizations that are apart from business or government. They are often nongovernmental organizations (NGOs). The range of civil societies is large and informal. Community organizations of all types— religious organizations, professional organizations, labor unions, and school organizations—are civil societies. The idea behind their inclusion is that they provide a strong voice of the people in democratic nations.

The United Nations especially recognizes the role of civil societies in sustainable development. This is a delicate issue for the United Nations because member states do not want to be circumvented by their civil societies, as noted previously.

The civil society movement holds great potential for the equitable component of sustainability. The major world organizations advocating for sustainable development recognize the need to bring in all groups of people, whether or not the government includes them. It is necessary to protect and evaluate ecosystem viability knowledge from these groups. Without this knowledge of ecosystem risks, other aspects of sustainability become difficult to implement. For example, the application of the precautionary principle, which determines if there is a irreparable threat to the systems of nature on which future life depends, is stymied without accurate and complete knowledge of the ecosystem. In many nations, however, there is resistance from incorporating the civil societies of that place.

The values of civil societies range over a wide spectrum and can change over time. They can be influenced by the economic and political power of the nation, or the nation can be at odds with the values of the civil society. International organizations like the ones discussed are organized around nations and, as such, are very careful when they engage civil societies. They do not want to threaten their member nations with inclusion of civil societies. Other international organizations in the private sector are often sought to finance or manage large developments with substantial environmental and economic impact. Their perspective on civil societies is one of less engagement. They do not want to interfere with their potential for profit or engage in the political affairs of the nation. As such their environmental impacts can be unchecked by civil societies, and these societies may be the only ones knowledgeable of the impacts. This develops the dynamic of civil societies in controversial environmental projects going to international bodies and circumventing their governments. To the extent that processes are in place that includes civil societies the equitable component of sustainability increases.

Governments are increasingly working with NGOs and business to achieve sustainability goals. The UN is actively seeking to make NGOs a part of dialogue on important global issues. The UN Economic and Social Counsel is particularly involved, holding meetings and hearings with NGOs through the commissions on sustainable development, the status of women, and population and development. The Millennium Development Goals program also actively seeks to involve NGOs.

The following UN projects seek to help civil society achieve sustainability goals:

- Non-Governmental Liaison Service

- U.N. Civil Society Web site (allows lay-user to learn about and navigate U.N. resources)

- United Nations Development Programme

- Economic and Social Counsel

- Community Commons

- Department of Public Information, NGO Section

- Millennium Development Goals Campaign Office

References

UN Global Compact. www.unglobalcompact.org/index.html.
UN, General Civil Society Hearings. www.un.org/ga/civilsocietyhearings/.
UN, Non-governmental Liaison Service,www.un-ngls.org/.

Social Capital

Social capital is a concept that places value on participation in social networks. Social contacts developed through social networks can affect the productivity of groups and individuals. One way that social capital can have value is through consensus building. Group consensus means that networks have shared interests and agree about how to accomplish certain goals. Social capital is often linked to the success of democracy and political involvement, and The World Bank defines social capital as the norms and networks that enable collective action. This goes to show that the ideas and principles of social capital are multidisciplinary.

Social capital also incorporates the influence of social networks; these can be tangible, such as ethnic enclaves that form with transnational immigration, or intangible, such as social networking sites that connect people around the globe. There is no widely held consensus on how to measure social capital. The measure of social capital is often a product of the type of networking that is quantitatively and qualitatively being measured. Some measures of social capital have to do with the size of a group, whereas another measure is the group's cohesiveness.

Social capital is concerned with the connections people make and the reputation and influence the resulting networks have. The importance of making connections is not limited to those among individuals. Connections across different networks and between groups also work to increase social capital. The quality of the connections—how deep or meaningful they are—can also have an impact on the cohesiveness and influence a group has. Social capital is useful in that it allows members of a network to access ideas, information, and resources through connectivity. There is also an element of reciprocity built into social capital networks. If one network provides resources to another network, the bonds are strengthened and the resources exchanged will eventually go in the other direction.

Despite the positive impact of social capital, it is important to note that strong internetwork ties can weaken the strength of intersocietal ties. This is especially true when the networks that are creating social capital are engaged in crime or out-group discrimination.

The term *social capital* was used by Jane Jacobs in her book, *The Death and Life of Great American Cities* and was popularized by Harvard sociologist Robert D. Putnam in his book, *Making Democracy Work*. Putnam defines social capital as the "features of social organization such as networks, norms, and social trust that facilitate coordination and cooperation for mutual benefit." The World Bank refers to social capital as "not just the sum of the institutions which underpin a society—it is the glue that holds them together."

References

DiMento, Joseph F. C., and Pamela Doughman, eds. 2007. *Climate Change: What It Means for Us, Our Children, and Our Grandchildren.* Cambridge, MA: MIT Press.

Field, John. 2006. *Social Capital.* Andover, UK: Routledge.

Jacobs, Jane. 1961. *The Death and Life of Great American Cities.* New York: Random House.

Putnam, Robert D. 1993. *Making Democracy Work.* Princeton, NJ: Princeton University Press.

Putman, Robert D. 2004. *Democracies in Flux: The Evolution of Social Capital in Contemporary Society.* London, UK: Oxford University Press.

Differentiated Leadership

Sustainability can mean different things depending on a country's economic and social conditions. In some countries, the need to develop is driven by basic human needs, and in others, consumption of resources far exceeds basic human needs. Sustainability may mean reduction in consumption for some, and increasing development without unsustainable methods of production in others. Leadership toward sustainability must mutually recognize different approaches for different conditions.

The idea of differentiated leadership roles tries to reconcile these differences without hypocrisy. For example, the United States uses far more of the world's resources per capita than any other nation on earth. Leadership for sustainable development in the United States must reduce consumption and waste. Countries like Haiti consume far less per capita, although they are much more densely populated. Leadership for sustainability there must develop new sources of food, energy, clothing, and basic shelter in ways that maintain and restore this island's decimated ecosystems. Another leadership challenge in developing countries is the problem of governmental corruption. A leading civil society organization, Transparency International, defines governmental corruption as "the abuse of entrusted power for private gain. It hurts everyone whose life, livelihood or happiness depends on the integrity of people in a position of authority."

ENVIRONMENTAL JUSTICE

Environmental Justice and Sustainability

Environmental justice refers to the disproportionate impact of environmental decisions, usually by race, class, income, or gender. It also incorporates ideas about fairness in access to environmental decision making. Environmental decisions become unjust when tainted by racism. Systems of nature on which future life depends can be irreparably damaged by environmentally unjust decision making.

Reference

"Executive Order: Federal Actions to Address Environmental Justice in Minority Populations and Low-Income Populations." www.epa.gov/Region2/ej/exec_order_12898.pdf.

African Slavery and the Cruelty of Lynching

U.S. slavery began in 1619 with the first official sale of Africans in Jamestown, Virginia. It officially ended 246 years later. Slaves were not allowed any human rights because they were considered property. Therefore slaves were denied the right to language, education, marriage, and possession of property. Slavery officially ended in 1863 when President Abraham Lincoln signed the Emancipation Proclamation. The period from 1865 to 1896, however, was as brutal as slavery. Even after 1896, African Americans were brutally treated with no legal recourse. The brutality of treatment was highlighted by lynchings, which were gross acts of inhumanity. Audience participation was encouraged. Consider this entry in the *Vicksburg Evening Post,* February 13, 1904.

When the two Negroes were captured, they were tied to trees and while funeral pyres were being prepared, they were forced to suffer the most fiendish tortures. The blacks were forced to hold out their hands while one finger at a time was chopped off. The fingers were distributed as souvenirs. The ears of the murders were cut off. Holbert was beaten severely, his skull fractured and one of his eyes, knocked out by a stick, hung by a shred from the socket. . . . The most excruciating form of punishment consisted of a large corkscrew in the hands of some of the mob. This instrument was bored into the flesh of the man and woman, in the arms, legs, and body, and then pulled out, the spirals tearing big pieces of raw, quivering flesh every time it was withdrawn.

Reference

Allen, James et al., 2000. *Without Sanctuary: Lynching Photography in America.* Santa Fe, NM: Twin Palm Publishers.

The Evidence of Environmental Injustice

In many cultures the race of people affects their power and degree of political and economic marginalization. In the United States the United Church of Christ researched and published a ground breaking study in 1987 that demonstrated that the more hazardous the waste, the more likely it was located in an African American community. The study showed that it was a 1 in 10,000 chance that this result was random, or a 99.9 percent degree of certainty. This study was one of the strong roots of the modern day environmental justice movement. In 2007, researchers revisited this study to see what had changed. From 1987 to 2007, three comprehensive reviews of existing research were completed. Each review found that environmentally disproportionate impacts were distributed by both race and socioeconomic status. The 2007 report referenced in this chapter indicated that the racial disparities are even greater than first measured as a result of more accurate measures. Geographic information systems, more accurate environmental reporting, and more engaged citizen activism all contribute to a more accurate assessment of environmental impact.

Hazardous wastes are kept in specially designed facilities called treatment, storage, and disposal facilities (TSDFs). The 2007 report found that the proportion of people of color estimated to be within three kilometers of TSDF to be disproportionate. Three kilometers is the distance used because it is commonly accepted as the distance in which adverse health impacts, property values, and quality of life impacts suffer. In the 2000 Census, there were 413 TSDFs. They were located in African American communities three times more than white communities. This is just one small dynamic of a much larger dynamic around waste and race. TSDFs are only one type of waste facility. There are many more and waste is accumulating in all of them. As the adverse impacts, environmental degradation, and declining property values about waste become more known, this dynamic may become more entrenched in land use patterns.

Waste is almost always increasing, and it must go somewhere. It is a direct threat to sustainability because it can overwhelm systems of nature on which future life depends. Racism in the structure of human habitation does not decrease waste. Knowledge of how waste is affecting a given ecosystem is vital to developing an accurate environmental baseline of that ecosystem. The equity component of sustainability includes those most affected by environmental degradation because that knowledge is important for systems of nature to survive for future generations. Waste decisions are just one aspect of environmental justice that affects sustainable development.

In 1992, the Environmental Protection Agency (EPA) concluded that racial minorities and low-income population experience higher than average exposures to (1) certain air pollutants, (2) hazardous waste facilities, (3) contaminated fish, and (4) agricultural pesticides. The EPA also concluded that there was insufficient data to say where these effects are "adverse," with the exception of childhood lead poisoning. Children of poor families are eight times more likely than children of high-income families to have lead poisoning and African American children are five times more than white children to be lead poisoned. In 1994, 64 empirical studies examined a wide range of environmental hazards, and 63 studies found disparities by race or income. When race and income were compared, race was found to be more important in 22 of 30 studies.

Farm Worker Exposure to Pesticides

Farm workers are often exposed to chemicals in the planting, growing, harvesting, production, and shipping of crops. Farm workers are often people who follow crops harvesting cycles and migrate from place to place. If they complain about illegal pesticide use, they are often retaliated against by losing their job and sometimes their homes. Farm workers are generally paid less, have very few health benefits, and do not receive typical municipal services. They are allowed to be paid less than the minimum wage by law because of a 1930s exception to the National Labor Relations Act that was designed to prevent former African American slaves from competing with white farmers. There are about 1.2 billion pounds of pesticides sold for $4.6 billion per year in the United States. Approximately 600 active chemical ingredients are combined with others to form about 35,000 different chemical formulations.

About 313,000 U.S. farm workers suffer from exposure-related illnesses and about 800–1,000 die. Approximately 90 percent of farm workers are people of color—mainly Chicano, followed by Puerto Rican, Caribbean blacks, African Americans, and indigenous peoples from Mexico. Some but not all farm workers are illegal immigrants. This makes it easy to retaliate against them for reporting pesticide abuses. Many protective measures are ineffective such as protective clothing. In the United States, ethyl parathion is the leading cause of farm worker poisonings. Generally, organochlorides create bad health effects over time and persist in the environment. Organophosphates do not persist, but they have strong effects on humans, as they are acutely toxic. Internationally, many nations allow any type of pesticide. Less developed nations often allow them if they can afford them.

References

Gilbert, Charlene, and Eli Quinn. 2000. *Homecoming: The Story of African American Farmers*. Boston: Beacon Press.

Moses, Marion. 1995. *Designer Poisons: How to Protect Your Health and Home From Toxic Pesticides*. San Francisco: The Pesticide Education Center.

Williams, Juan. 2006. *Black Farmers in America*. Lexington: University of Kentucky Press.

EXPOSURE TO CONTAMINATED FISH

Most studies of fish consumption overlook subsistence fishers. These are groups of people who rely on fish as a food source. Many native people rely on fish for food and as an important cultural value. Many treaties include provisions for free access to ample fish stocks.

Not only are many fish stocks depleted, they are highly contaminated. There is no longer any river or lake in the lower 48 U.S. states where it is completely safe to eat the fish because of the chemicals from pollution. If a small amount from a part of the fish that does not accumulate the chemical is eaten, it could be safe. Many cultural groups rely on fish to survive and eat most of the fish, including the parts that accumulate pollutants. These groups are disproportionally impacted by the environmental decisions that allow the lakes and rivers, and the fish, to become contaminated.

TOXIC RELEASE INVENTORY FACILITIES

The basis of the national right to know law is the Toxic Release Inventory (TRI). The TRI is a federal list, or inventory, of the largest emitters of regulated chemicals. Most of the data are self-reported by industry. It is a good community-organizing device and a way to understand a community's particular perception of risk because it is available to the public. It lists about 75,000 industrial facilities and reports releases of about 600 chemicals. Industrial facilities with toxic release emissions outnumber hazardous waste facilities by almost 40 to 1.

TRI has been an environmental battleground because of the community-based environmental justice issues. The 2006 Bush Administration tried to undermine the program by deleting reports from more than 500 industrial facilities that release up to 2,000 pounds of chemicals every year. They also deleted reports from almost 2,000 industrial facilities that manage up to 500 pounds of destructive chemicals such as lead and mercury. They tried to change the reporting requirement from once a year to once every two years. These and similar efforts to hide the environmental and public health impacts of pollution by industry were met with strong resistance.

TRI facilities can be used for an analysis of the distribution of these facilities by race and income of the communities. Most research about these distributions demonstrates that race matters most in terms of African Americans receiving the most exposure, and that class also matters. Working class neighborhoods, not poor ones, receive the most exposure. This is true even when controlling for other variables such as income, education, and age.

References

Agyeman, Julian, Peter Cole, Delay Haluza,-and Pat O'Riley, eds. 2009. *Speaking for Ourselves: Environmental Justice in Canada.* Vancouver: University of British Columbia Press.

Brooks, Nancy, and Rajiv Sethi. "The Distribution of Pollution: Community Characteristics and Exposure to Air Toxics." *Journal of Environmental Economics and Management* 32 (1997):233, 243–46.

Mohai, Paul, and Robin Saha. "Reassessing Racial and Socioeconomic Disparities in Environmental Justice Research." Demography 43 (2006):383.

U.S. Government Accountability Office, Environmental Right-to-know, www.goa. gov/new.items/d08115.pdf.

AIR TOXICS EXPOSURES

The differential exposure to air pollutants by race and income is an environmental justice concern. Industrial facilities; proximity to highways, ports, trains, and bus depots; and medical incinerators all increase the amount of air pollutant in an area. Studies have started to measure environmental impacts by how they affect the people who live there. This is inclusive of cumulative risks measured by cancer risks. One study examined the Southern California Air Basin and 148 toxics hazardous air pollutants (HAPs). The study examined median cancer risk by race/ethnicity. Even controlling for income, these risks persisted. For person of color, covering African American, Latino, Asian, and Pacific Islander, risk of cancer is almost one in three as a result of air toxics exposures. For an Angelo, that risk is about one in seven.

Reference

Morello-Frosch, Rachel, Manuel Pastor, and James Saad. "EJ and Southern California's Riskscape: The Distribution of Air Toxics Exposures and Health Risks among Diverse Communities." *Urban Affairs Review* 36 (2001):551.

Disparities in Environmental Enforcement and Cleanup Priorities

This aspect of environmental justice examines environmental decisions around the enforcement of environmental laws, and decisions around the priority given to cleaning up environmentally degraded sites. A 1992 National Law Journal Study found that the U.S. EPA gave white communities faster responses, more thorough results, and higher penalties for the same environmental infractions than for communities of color. This same report concluded that communities of color did not get cleaned up as fast and did not make the National Priority List of Superfund cleanup sites as frequently as white communities. Another study found that in communities made up of more than 25 percent of people of color, there is almost nine times the cumulative rate of exposure to hazardous materials than in predominantly white communities.

The fair and equal enforcement of environmental law is important to sustainable development. It is often the case of politically marginalized groups that the environment is so degraded, that it could affect systems on which future life depends. Most environmental law is not adequate for sustainability right now because it does not reach all past and present environmental impacts accurately. The communities with the most environmental impacts may know more about them than environmental regulators. The equity component of sustainability would include those communities living in environmental degradation in order to gain accurate knowledge of the environment in a way that is useful for sustainable development.

The fury unleashed by the report by those against environmental justice only underscored the strength and depth of the institutional racism that permeates U.S. environmental decision making. The relationship between toxic waste and race was repeatedly challenged and revisited. Environmental justice, like any other kind of justice, relies on truth. Unlike abstract forms of justice, the environment provides solemn testimony to past and present injustices. By analyzing the depth of environmental degradation in every community, the environmental reparations necessary for a sustainable community can be decided.

INSTITUTIONALIZING ENVIRONMENTAL JUSTICE: EXECUTIVE ORDER 12898

In 1994, President Clinton signed Executive Order 12898. An executive order is not law, but an order from the chief commanding officer to all federal agencies. This Environmental Justice Executive Order required most federal agencies to examine disproportionate impacts in their decisions by both race and class, encouraged the participation of minority and low-income people in environmental decisions, and started the Intergovernmental Working Group. Environmental justice is referred to as an aspiration or goal, not as a problem. Many states have copied this executive order to create state level executive orders from their governor.

Environmental equity, environmental justice, and environmental racism are all terms that have come to the center of urban, state, national, and international environmental decision making. Citizens, environmental advocates and their organizations, big and small industries, elected officials, and government policymakers are all now aware of these terms. The dynamics of cumulative emissions and bioaccumulative impacts, combined with continued population growth, increasing waste and decreasing waste sites, and by-products of technology all increase the concern and study of the distribution of environmental benefits and burdens. Any bioregionally based approach to sustainability will sharpen the focus of fairness in environmental policy, law, and personal practices. The relationship of race, class, and gender to environmental policy and law in the United States historically, currently, and prospectively is of overarching topical importance.

UNITED CHURCH OF CHRIST (UCC) REPORT

Without the groundwork laid out by the UCC report, it is unlikely U.S. urban environmentalism would exist. Without urban environmentalism, the U.S. environmental movement would stagnate and U.S. environmental policy would remain ineffective in the face of accumulating chemical emissions and broadening exposure vectors to all parts of the U.S. population. The large disparity in the location of waste facilities with regard to race was, and is, indisputable despite scores of attempts to prove otherwise. By focusing on the actual place of the waste, and by letting those communities around the waste speak for themselves, a whole new dynamic of urban environmental policy began. Before the UCC report, it was difficult to assess any demographic characteristic with any type of environmental impacts. U.S. environmentalism is distinctly antiurban; however, most of the pollution is in cities, where most people of color live, work, and play. This report laid the basis for countless syllabi, state and federal legislation and rules, and many other reports. One big implication for the UCC Report and the need to speak out was an explosion of different methodologies. Every time geographic information systems evolve in technological refinement and are applied to environmental decisions, there is only greater evidence of disproportionate impacts, usually by race. When the TRI was made public, data not only about the amount and kind of chemical but also which communities were getting it was made available. Old ways of making public policy that relied on slow, inaccurate, and incomplete case study methodologies were directly challenged by technology, better environmental data, and the need to know the environmental truth of any one place. As environmental impacts accumulate and public concern for sustainability increases, place study methodologies develop for application in urban areas.

Reference

Bullard, Robert. 2000. *Dumping in Dixie: Race, Class, and Environmental Quality.* New York: Westview Press.

Neal, Ruth, and April Allen. *Environmental Justice: An Annotated Bibliography.* www.ejrc. cau.edu/annbib.html.

GLOBALIZATION: INCREASING INEQUITY

Globalization refers to the ever-increasing network of connections between individuals and organizations that is made possible by technology and transportation. These networks include information technologies and other communications that facilitate the transfer of capital and goods, and the redistribution of labor around the globe. Globalization can mean ever-widening markets, as governments agree to remove barriers to trade and immigration between them. Markets have been good vehicles for increasing wealth, but their growth has also accompanied alarming trends in the rate of climate change and other environmental deterioration, as well as growth in poverty. The energy that fuels our economies depends on combustion of fossil fuels that accelerate the phenomenon of global warming. Globalization of our markets and communications contributed to an increase in the production and consumption of goods that are exorbitant consumers of resources. The consequences of these changes have hurt the world's vulnerable people and communities the most. These two trends continue to worsen rapidly, and on a global scale. In our contemporary world, economies are connected by many links including global communications networks, transportation networks, and banking and finance networks, creating ever-larger global marketplaces in which goods and services are traded. This trade prompts ever-growing demands on our ecosystems that provide the resources and energy necessary to provide for our human enterprises and communities. Multinational corporations bigger and more powerful than many developing nations seek the cheapest natural resources in the most unregulated environments with the lowest labor cost. The impacts on ecosystems and other aspects of the environment are not a consideration for them and may not be known by the developing nation.

Globalization creates new markets and wealth. These new markets are often very lucrative for some people in the world. Nonetheless, globalization can also be the cause of widespread suffering, disorder, and unrest. Globalization has also served as a catalyst for change and social movements. As cultures interact, advocates for equal treatment for all global citizens may be spurred to action. Globalization can bring immense change often benefiting those most in need. As the term implies, globalization touches the lives of people all over the globe.

Communications have increased exponentially using satellite and microwave technologies together with digitization of data. Computers have made management of these technologies possible on a grand scale. These communications technologies by themselves take a toll on resource and energy use. The average laptop uses 45 tons of raw materials and is rarely recycled. Many of these technologies use energy even when they are not in use, prompting the name "energy vampires." They have also been used for business and commercial purposes. Capital is readily transferable using these communications technologies. Producers now sell to a variety of buyers that were not reachable in the ordinary sales transaction without communications technology and computers. This has increased demand for products. In the production of new products, raw materials can be purchased and transported for assembly and resale globally. The availability of transportation has also made labor itself a global commodity, with businesses moving toward the least costs of production including cheap labor. Traditional governmental restrictions on the movement of goods between marketplaces have been removed through trade agreements. In these ways, the marketplace for goods and services has been greatly expanded. The resultant growth in markets and transnational corporations has dwarfed many national economies.

Transportation is also global and much quicker than ever before, but the dependence of transportation companies on oil and gas has made this network unsustainable. Wasteful and inefficient types of capitalism can allow globalization to increase poverty and environmental degradation. The assumptions around the energy cost of transportation and the ability to transport goods may affect the type of globalization that we see in the future.

If businesses thus linked into a global economy are not producing in a sustainable way, globalization threatens to intensify pressures on ecosystems, resources, and energy in ways that spiral ever more quickly toward ecosystem collapses. Meanwhile, the social and environmental dangers of globalization may be masked by the phenomenon of externalization of costs until the harms are disastrous. The ability to harness the power of markets without triggering collapse of critical ecosystem functions is another area of intense controversy, even though increasing costs of transportation of goods may undercut the salience of global markets.

Globalization is a complex idea that encompasses the integration of economic, political, and cultural system across the globe. There are conflicting ideas about whether globalization is a positive force that includes economic growth, prosperity, and ideas about democratic freedom, or whether it fosters change that includes environmental devastation, exploitation of the developing world, and suppression of human rights.

There has been a recent increase in the pace and scope of contact and ties between human societies across the world. This has been spurred by technological advances in communication and transportation. National borders are increasingly porous, allowing a diverse group of people to interact. Because of the effects of globalization, ideas and cultures circulate more freely. This mixing of ideas can have a positive impact; however, these changes are also uprooting old ways of life and threaten cultures that have previously been insulated from outside impact. Knowledge of ecosystem services can be exploited before its true value is realized. Impoverished nations may be dumping grounds for dangerous products that erode public health and the environment, for example, pesticides that are illegal in developed nations. The threats posed to sustainable development by globalization of wasteful capitalism can be countered by the promises of global cooperation and sustainable technology transfer. *See also* **Volume 3, Chapter 5: Future Directions and Emerging Trends; Appendix B: Millennium Development Goals.**

PUBLIC PARTICIPATION: SUSTAINABILITY AS A PROCESS OF INCLUSION IN ENVIRONMENTAL DECISION MAKING

Under equitable aspects of sustainable decision making, poverty is a major issue. It impairs robust public participation and denies decision makers a source of onsite environmental information. Sustainability may create new labor demand for onsite cleanup and environmental monitoring and, in this manner, transact social capital.

An equity component of sustainability is inclusion. Public participation in the decisions that most affect a population is a fundamental assumption in many democracies. Public participation can take many forms. Many of the 1970s environmental laws, such as the Clean Air Act and Clean Water Act, had mandatory public participation requirements. Public participation can often take meaningless forms that simply go through the steps of the law after a decision is already made. Public participation is not cost free.

Reference

Cole, Luke. 2001. *From the Ground Up: Environmental Racism and the Rise of the Environmental Justice Movement.* New York: New York University Press.

Public Participation in Environmental Decision Making

Conflicts arise as communities strive to participate in environmental decision making. Communities are often not included into early stages of dialogue concerning the use of community resources when decision making is most meaningful. To be heard, communities are often forced to seize opportunities to participate in controversial ways, in the absence of a required public process such as the Aarhus Convention. In the United States, these efforts are being developed through the environmental justice movement.

One of the traditional methods of resolving environmental disputes in countries that rely on common and civil law is the use of courts. In the United States, a common law country, the person bringing the lawsuit, or filing the complaint in court, has the burden of proving the case in court before a judge and against a defendant. The burden of proof is a slippery concept in any given case, and courts have been shown to be politically influenced in environmental decision making. Courts are not accessible to everyone in most nations. Individuals must have standing, must state and prove a limited list of cases of actions, and must be injured in fact. In some states, such as Oregon, courts do not recognize causes of action against certain protected natural resource industries such as logging. Many environmental concerns of residents are shunted into meaningless land use forums that do not address environmental issues. Courts are based on an adversarial two-party decision backed by the law. Many environmental and sustainable decisions have many more parties of interest. They may prefer nonadversarial collaboration over the long term as opposed to adversarial, quick decisions. Many plaintiffs cannot afford the scientific costs and risks of bringing such a lawsuit, proving harm in fact from a given environmental cause, often called toxic torts.

Public Participation under Sustainability

There are two opposing decision-making models in the United States and many other nations when environmental decisions are made. The traditional and far more common approach is pluralism. This refers to policy development and public participation as a political marketplace where different interest groups struggle and compete for scarce resources and political victories.

Pluralism has five characteristics. First, it requires forums for bargains, exchange, and conflict resolution. These can be local land use meetings or a community meeting on an environmental impact statement. Second, the decision maker, usually an agency head, city council, or a judge, aggregates, or combines, all the preferences of the represented stakeholder. Sometimes a single stakeholder may capture the entire process, which is called capture. It is used to describe industry capture of EPA or state environmental agency decisions. If the process is exclusionary, it increases the change of solo stakeholder capture, and the process loses some of its legitimacy. In sad but accurate reality, the decision maker has already expended considerable energy, resources, and expertise on the decision (e.g., an environmental permit) and will not be likely to amend it or deny reflecting public concern. Many criticize the pluralistic nature of the current process because the timing of public participation allows industrial and commercial stakeholders to capture the process of dialogue with the agency or city. Most have described pluralism as a crude preference tally at best. The only unpaid stakeholder in these forums is often the community resident. Third, pluralism is characterized by unequal power distribution among stakeholders. The ability to be part of decision making is centralized between one or two stakeholders. Fourth, pluralism is also characterized by large differences in access to important information. A public hearing may take testimony but not answer any questions. A fifth characteristic of pluralism is that stakeholders vote or act to decide based on individual self interests, not based on the collective good, public health, safety, or ecological impacts unless that is their particular stakeholder group. Because groups representing these interests are excluded or marginalized using many pluralistic policy practices, their environmental decisions may not hold up to scrutiny when viewed from a sustainability perspective. In contrast, sustainability requires deliberative dialogues, where the common interest is the common goals. It requires engaged and knowledgeable participants and engaging in rational, nonadversarial discourse about what outcomes best serve the common good of the community. The common good is not merely aggregated and bargained preferences among a small and select group of participants. Participants, that is, stakeholders, are expected to set aside and revise personal preferences in the interest of finding shared values. In the deliberative process, stakeholders find common good through discourse and shared capacity building, as opposed to discovering it through preexisting preferences. When applied to land use, some of these processes are called visioning.

If sustainability requires a personal commitment to act in certain ways and make substantial behavioral changes, as some maintain, then the voluntary and willing buy-in of everyone may be necessary. This level of personal commitment is often beyond the coercive power of the state, which is often very limited and is used to protect the status quo. In many nations, this amount of intrusion by government is seen as a violation of liberty. Sustainability will require some level of personal commitment, and individuals will behave in ways that are sustainable when they perceive they are being treated fairly. Humans are considered more than a life form that is part of the ecological system; humans are controlling actors who can transcend their own existence and alter current unsustainable practices.

Barriers to Participation

A major concern that haunts U.S. environmental decision making around public participation is that industry fears that environmental groups use the mandatory public engagement processes to find potential plaintiffs for lawsuits. The general rule is that the administrative processes must be exhausted before a lawsuit can be filed. The environmental laws with citizen suit provisions often require a signed affidavit from a citizen or citizens group. Industry is reluctant to participate any more than necessary if it is going end up in court. Unless the potential for profits is very high, many industries shy away when public participation around their environmental impacts is closely analyzed.

PROCEDURAL JUSTICE: BASIC THEORETICAL COMPONENTS OF FAIRNESS

Procedural justice is a function of the manner in which a decision is made—the fairness of the decision-making process—rather than on its outcome. In law, procedural fairness is a matter of adequate notice and opportunity to be heard. In terms of equity, procedural justice determines that a particular judgment is fair because of the perceived fairness of procedures leading to that outcome. There is a strong value in the United States around the idea that people have a say about that which personally affects them. What happens when a violation of procedure occurs to get to a result that is distributively unjust? Although law may require compliance with the particular decision, it may be minimal. When personal consumption patterns may have to change to achieve sustainability, minimal compliance may not be enough. In contrast, the new value of inclusion resonates with the procedural fairness concept because participation is part of a fair procedure. This includes equal access to information, advance notice of meetings and decisions, and measurement of all environmental impacts.

A major part of the equity component of the definitions of sustainability is procedural fairness. This generally refers to the role of government regarding both citizens and industry. For sustainability to be implemented, it is easier if it is perceived as "fair." In most applications of procedural fairness, the environment itself is not represented. Another component of equitable aspects of sustainability is its emphasis on inclusion. Inclusion brings with it a multistakeholder perspective in the context of rapidly expanding environmental information. It can also affect the procedural aspects of the strongly held value of "fairness." *See also* **Volume 2, Chapter 5: Equity.**

OPPORTUNITY TO BE HEARD: REPRESENTATION OR DIRECT PARTICIPATION

In terms of procedural fairness, representation refers to the extent that parties to a dispute believe they had the opportunity to take part in the decision-making process. It is an implicit U.S. value. Democratic governance allows for a say in those decisions that affect our public health, safety and welfare, and environment. The more personal a decision, the more politically salient it is, as in children, property, contracts, crime, environment, then the more communities feel strongly about having a say. In such a large paradigm change as sustainability, communities will welcome this aspect of inclusionary dialogues. Having a say, however, does not mean getting one's way in a particular decision. This can be very frustrating to newly included stakeholders such as many communities in developing new practices and policies of sustainability. In land use and local environmental issues, there is little right to representation to anyone but property owners and contiguous property owners.

Fair representation allows for greater compliance by society of rules, theoretically yielding more order. Law creates and enforces order in society, a strong value in most nations. The assumption is that people will comply with the rules even if the outcome is unfavorable to them because the process allowed them to have a say.

CONSISTENCY OF TREATMENT AND POLICY

For a process to be considered "fair," it needs to be consistent. Everybody in similar circumstances needs to be treated the same. The U.S. Constitution and some state constitutions assert equal protection of the laws, and its equal incorporation, as a condition of statehood. Consistency is also an aspect of fairness evidenced by the treatment of people over time. This aspect of the strongly held value of "fairness" could be a challenge for sustainable decision making, especially at first. Many implementing decisions of a policy of sustainability could seem unfair because they may require changes in other, long-held values like private property. Given the lack of knowledge about the environment in terms of current conditions of the ecology and how to make it and keep it sustainable, there are likely to be a range of decisions. The spotty current use of the precautionary principle in land use may also create a lack of consistency in land use decision making. This has large financial implications for land or natural resource-based economic development. Property transactions, trusts, estates, contracts, and economic planning all prefer a stable, predictable, and consistent approach to real property and land use control. Many land owning private citizens also want to be assured of the sanctity of their home. Many sustainability advocates, however, point out that a policy or value structure can be consistent, while being consistently wrong. Environmentally based sustainability advocates also point out that species survival usually depends on adaptability to changes in climate over time. This is a powerful and unavoidable controversy, pitting powerful vested stakeholders against emerging local and global environmental realities.

IMPARTIALITY

This part of traditional procedural fairness means legal authorities suppress any biases they have about the parties or the outcome of the dispute. Personal characteristics affecting process or outcome, either positively or negatively, are ignored. Race, gender, income, social class, religion, marital status, age, or an aspect of a person is to be ignored in judging the facts of a given decision. The inclusionary aspect of sustainability would bring in urban populations of disenfranchised people in environmentally degraded places. Difficult and controversial issues of the past may pull current definitions of environmentalism as practiced in sustainability policies into more public health and ecosystem decisions. Actually proving bias is very difficult against any one person and much harder against an institutional stakeholder like a city or a manufacturing plant. Greater knowledge of environmental impacts, however, may result in patterns of decisions that are based on partial decision makers. That is, cities and their land use decisions and environmental agencies and their regulatory decisions made many decisions in the past based on bias.

APPEALABILITY OF HARMFUL DECISIONS

This aspect of procedural fairness means that there are higher level authorities to which one can appeal the decision. Some nations have unlimited appeals if one can afford them. Some nations have no appeal and sometimes no trial. Under sustainability, decisions would have to be correctable if they were harmful to people or to the environment. To determine whether they were harming people or the environment more, environmental monitoring and much more robust reporting of environmental impacts are required. The correctability aspect of procedural fairness would be expanded under sustainability because of the lack of current knowledge, continuing growth in the application and development of environmental mitigation strategies and techniques, and the need to revisit past decisions that are no longer permissible.

The next step in traditional public participation planning for environmental decisions is to assess the level of controversy around a proposed development. This step is important because sometimes the level of controversy can be significant enough to require a full environmental impact assessment instead of just an environmental assessment. There can be many types of controversies in a given environmental decision. Many controversies are increasingly about inclusion, and this will be the case under sustainable decision-making regimes. Other controversies at this stage are generally about access to information or sometimes the accuracy of information. Under sustainability, the precautionary principle of slowing development to see whether there would be any irreversible impacts on natural resources and the environment would be applied. Current information controversies at the public participation planning stage of sustainable decision making may therefore require much more time and information. Nonetheless, some estimate of controversy is made. When assessing controversy at this stage, some basic questions are asked and sometimes answered. Is there any similar prior controversy on the same issue? This question is asked because it could be a sign of a large national or statewide environmental controversy. The planning team examines court cases, government documents, and other sources to make this evaluation. For example, if the decision is about the environmental impact of logging old growth forests, then the team would uncover a large controversy with many court cases, protests, and passionate participants. Another area the planning team would examine is whether the proposed decision is related to any other major issues occurring at the site. This is currently narrowly construed, but under sustainability would have to be expanded. Currently the inquiry is limited to legal claims, pollution liabilities, historical and cultural preservation, and endangered species. Under sustainability this inquiry would be expanded to include the ecology of the place and the history of land use at the site. A last set of questions the planning team asks is how significant the particular environmental proposal is to them. The assumption is that insignificant issues are probably noncontroversial, and significant issues are or can become controversial.

After an assessment of the level of controversy is made, the public participation planning team usually begins to assess the minimum level of participation required. There are about five type of roles assigned to participants. They can be observers only. They can be allowed to be informed. Second, people can be allowed to make comments, or to be heard. Third, people can be technical reviewers, which can exert strong influence on narrowing the process. Fourth, an active participant influences and shapes the decision, but does not make it. Last, a codecision maker or decision maker must agree to the decision and make it. Unfortunately, the usual focus on the process is how to achieve minimal compliance with the laws and rules, not achieve maximum meaningful involvement. Sustainability decision making may expand the roles of participation here.

How to Develop a Public Participation Plan

Although most environmental and sustainability issues are unique to a given place, when decisions need to be made, a plan is required to facilitate meaningful public participation. After a potential project with environmental impacts is proposed, the first question is who needs to be on the public participation planning team. This can be very political. Sometimes people are appointed to such a team in order to provide outreach to underrepresented groups. Usually the public participation plan team is composed of agency officials. The next set of questions is what are the issues and who are the stakeholders for this particular decision. This can be a difficult decision. After that, many public participation plans assess the level of controversy. This step is important because a high level of controversy can indicate the need for a complete environmental impact assessment. If there is controversy, the question becomes how to prepare for it. Then, for each step in the decision-making process, members of the public participation planning team asks what they want to accomplish with the stakeholders. They try to develop public participation objectives, usually including notice, information transparency, and cost goals. In developing these goals, the team asks what the stakeholders effectively need to know, and also what the team needs to learn from the stakeholders. Special circumstances such as language issues, disability access issues, and cultural issues are analyzed. These factors will affect the selection of public participation techniques. After the selection of which public participation techniques are deemed appropriate, the team prepares the plan by laying out objectives, goals, and a schedule.

The traditional model of public participation can be developed in either a pluralistic or deliberative manner, and is thus suitable for sustainability decision making if capacity building of all stakeholders is part of the process. The full blown model is very time consuming and increases the risk the project may be denied.

Who Needs to Be Included in Public Participation Planning?

There are three sets of general stakeholder categories: industry, government, and community. There can be many other stakeholders to a given decision about the ecology. Environmental, religious, educational, and professional organizations may also be stakeholders. Some stakeholders prefer to be silent and not participate, which is a problem for sustainability. Another challenge for sustainability is how to achieve representation of future generations. One response is that the assumption is that, by taking care of the environment and natural resources properly with policies of sustainability, future generations will be taken of.

Government is usually the convening stakeholder and controls who can be included. Beyond that, the team needs to consider people, business, and other governmental units that will be impacted by the decision or by open discussion of the topic. From this group the team then asks who can be called on to assist with the public participation effort. Including people with special expertise needs to be considered. This can be a thorny point because communities often want an independent scientific evaluation, but government and industry have their own scientists and seldom pay communities to hire their own. Scientists are often challenged in this context as not being independent, or not contributing meaningful knowledge to the process fast enough to serve the values of profit and economic growth. For sustainability, this aspect of the traditional process would need to be enlarged, with more independent scientists and the amount of time necessary to accurately determine whether there would any irreparable damage to public health or the environment, as well as to examine ways to mitigate these impacts. In France, universities sponsor community science shops to help address community concerns in a scientific manner. Grassroots science is sorely needed to help develop the environmental baselines needed for sustainability. Writers, graphic artists, computer technicians, and telecommunications services may also be required and can be costly in small or remote areas. Last, individuals should be included in a public participation panel who will lend credibility to the decisions. These can be church leaders, environmental organizations, and others.

Adverse Impact Analysis: Special Circumstances in Environmental Decision Making

Environmental public participation planning teams examine six special circumstances. First, are there any cultural or ethnic sensitivities or disproportionate impacts? Second, are any national interests at stake? Third, how far are stakeholders from the site and from any decision-making centers (e.g., a state capital)? Fourth, is the issue politically connected to other issues? Fifth, will some stakeholders be outraged by the decisions made, or are active and violent protests anticipated? And sixth, how likely is it that important politicians will have strong positions on these issues? With all these special circumstances, the question is how much impact they will have on the decision in front of the planning team. Under a sustainability decision-making regime, it is likely that special circumstance will be expanded to include time for actual environmental assessments and capacity building for better inclusion.

References

Allen, Patricia. 2004. *Together at the Table: Sustainability and Sustenance in the American Agrifood System.* University Park: Pennsylvania State University Press.

Bowler, I. et al. 2002. *The Sustainability of Rural Systems: Geographical Interpretations.* New York: Springer.

Cutter, Susan L. 2006. *Hazards, Vulnerability and Environmental Justice.* London, UK: Sterling, VA: Earthscan.

Robinson, Guy M. 2008. *Sustainable Rural Systems: Sustainable Agriculture and Rural Communities.* Aldershot, UK: Ashgate Publishing.

Stakeholder Involvement

There are generally several major stakeholder groups to consider in any community dialogue. Industry is often represented by businesses of differing size and can include trade associations. Community is often represented by civil society organizations that are mobilized to be interested and active. Labor organizations are participants who speak to issues of economics, as well as indoor pollution, waste, and other types of hazards that affect workplaces. Religious or faith-based organizations can provide important access to participation in ethnic communities. Representing the environment itself can be challenging and is often assumed to be the task of environmental activist groups. Finally, representing the interests of future generations is sometimes assumed to be the role of women, as caregivers. In other situations, indigenous people's organizations speak to the issues of environment and future generations. None of these assumptions should take the place of a broadly inclusive process.

NOTICE OF MEETINGS

People generally want notice of the meeting well before the actual meeting time. They want it at a convenient time and place. These meetings can be difficult for low- and middle-income families to attend. In most U.S. cities, only property owners get any notice at all, leaving renters uninformed. More low-income people rent than own property in most cities. The government is generally legally responsible for the notice. Most legal notices are printed in the back of the classifieds of the local or closest newspaper. They are generally in English, and sometimes in Spanish.

Actual notice of participation opportunities to stakeholders like renters, environmental activists, students, and others is limited. Many low-income people cannot attend these meetings because they cannot leave work without losing pay. Others have to make expensive and timely coping decisions around childcare and transportation just to attend a meeting. Most of the time, the best notice anyone gets of local environmental decisions is colorable notice, which is very weak notice. Basically it is information that is disseminated if requested or if you know where to look for it. Sometimes it is in the small print in the legal advertisements under the classified sections of local newspapers. This "notice" is information disseminated by local, state, and federal environmental agencies, and sometimes by industry This can be a confusing and fruitless process. The information provided can be overly scientific, can make false promises, and may even be factually wrong.

The concern of many stakeholders about giving too much notice is that it will create resistance to the project at hand. This is an indication of the strength of the value of growth and economic development. Local land use forums and environmental controversies are effectively foreclosed from meaningful consideration when decisions about economic development are first made secretly, and the public participation becomes merely pro forma afterwards. This also can cause erosion of trust in local government.

Notice is not cost free, nor is lack of notice without the impacts. For truly meaningful participation, the information has to be understood, and the community must have the capacity to engage in the environmental facts and the environmental impacts of a given proposal. All environmental information should be transparent, that is, clear and available to all parties. The problem with this approach in contemporary land and environmental decision-making forums is that it is very time consuming. This indecision can scare away potential economic redevelopment and growth. There are many challenges. The ability of a community to have the time and expertise to engage in a public process is uneven. The are unanswered questions about who pays for the this capacity building in the community, for outside expertise, for the mailings and other telecommunications costs, and for childcare and lost wages.

CAPACITY BUILDING

Stakeholders are usually not equal in terms of knowledge about the issues they face. Part of the task of creating the ability to have mutual discourse is the requirement of equalizing knowledge and the ability to understand data and consequences. Increasing the ability of stakeholders to understand the knowledge essential to their decision is often called capacity building.

ISSUE MANAGEMENT

The next step in developing a public participation plan is to develop an issue management plan. There are some general categories of issues that the public participation team must resolve. For the issue to be resolved successfully in a particular environmental decision, the team must decide if there other, preemptive policy decisions that must be made before the respective issue can be resolved. Decisions about what informational materials are needed are made here. Identifying key points of information exchange is important for public participation planning. Information used to define the problem, to establish the evaluation criteria, to identify alternatives, to evaluate alternatives, and ultimately to choose a course of action must be phased and divided into information going to the public and information coming from the public. The emphasis is usually on information going out to different stakeholders—communities, stockholders, banks, and government regulatory agencies. Under sustainability, the emphasis would have to increase information about the land use landscape and the public. In complex, multistakeholder environmental decision making, as in sustainability, this can be a hard question to answer. Science is often unable to answer specific questions about a particular environmental impact quickly enough under current models. If the precautionary principle is applied, however, stopping development until there is no danger of irreversible damage to natural resources may require giving scientists enough time to study the potential for this to happen. This is another area where sustainability will require more time to decide because of equitable concerns about inclusion, and in order to measure real environmental impact potentials. Public participation planners need to know if any studies need to be completed and who exactly is responsible for this process. Sometimes other actions may be needed, such as advance planning on how to get actual notice to all stakeholders.

A MODEL PARTICIPATION STRATEGY

Inclusion, which translates roughly into public participation, is a major part of the equity component of sustainability. Public participation in terms of notice and ability to comment is a traditional part of U.S. environmental decision making. Many have criticized this part as too little, too late for sustainability. Not all pollution is known or monitored, and it may be accumulating in population centers. Currently, more inclusionary model plans and idea about public participation are rapidly developing. This helps lay the foundation for implementing parts of the inclusionary aspect of sustainability.

In most model plans and emerging participation and involvement ideas, the goal is to explicitly encourage public involvement. One way to do this is to stop the process until a minimum number of a people, institutions, or stakeholders have been consulted with. Another way is to make all the information transparent, easy to understand, and freely available. Government stakeholders need to cosponsor outreach programs with community organizations of all types. When doing so, they can be equal partners in planning the agenda, making decisions, if any, and establishing goals. Communities often need to be educated and have free engagement with independent scientific expertise. Public involvement tends to increase when there is equal power-sharing among all stakeholders.

Some of the planning to encourage participation includes the basics like logistics. Handicapped accessibility, bathrooms, water, childcare, and payment for participation and/or participation costs are all important and sometimes expensive issues. With current technology, it is possible to "meet" without the necessity of travel for face-to-face communications.

References

Beierle, Thomas C., and Jerry Cayford. 2002. *Democracy in Practice: Public Participation in Environmental Decisions.* Washington, DC: RFF Press.

Blewitt, John. 2008. *Community Development, Empowerment and Sustainable Development.* Devon, UK: Green Books.

Dietz, Thomas, and Paul C. Stern. 2008. *Public Participation in Environmental Assessment and Decision Making.* Washington, DC: National Academies Press.

Kasemir, Bernd, and Jill Jager. 2003. *Public Participation in Sustainability Science: A Handbook.* Cambridge, UK: Cambridge University Press.

Paehlke, Robert. 2008. *Democracy's Dilemma Environment, Social Equity, and the Global Economy.* Cambridge, MA: MIT Press.

Tilbury, Daniella, and David Wortman. 2004. *Engaging People in Sustainability.* Gland, Switzerland: International Union for Conservation of Nature and Natural Resources.

FACILITATION

Facilitation is the process of coordinating meetings and bringing together multistakeholder groups in a variety of decision-making models, from none to consensus. Some U.S. federal agencies use facilitators in environmental decision making. There are private consulting groups, such as RESOLVE INC, that also provide facilitators. This is to be distinguished from Alternative Dispute Resolution, which is limited to two or three parties in an adversarial posture. Facilitators are valuable in most sustainability situations with inclusionary dialogue. They are trained in information and communication skills and must show no bias. They are seldom any part of the decision making. They can be expensive and are not used in most instances today.

Reference

Kaner, Sam et al. 2007. *Facilitator's Guide to Participatory Decisionmaking.* Hoboken, NJ: Jossey–Bass.

Current Public Participation Tools in U.S. Environmental Planning

THE U.S. ENVIRONMENTAL PROTECTION AGENCY: PUBLIC INVOLVEMENT POLICY OVERVIEW

The U.S. EPA has a strong commitment to legally required public participation. Depending on the environmental goals of the president and Congress, it tries to strengthen those legal requirements. With the rise of public access to more environmental information, many agencies are challenged to meet community demands for more involvement and more power. EPA underscores the need for earlier and more meaningful public involvement. It wants to ensure that environmental decisions are made with an understanding of the interests and concerns of affected people. It promotes more and more techniques to create opportunities for public involvement with EPA decisions. The EPA applies these procedures to many of its operations, such as making rules for environmental regulation, permit issuance, renewal, modification, selection of plans for cleanup of hazardous waste sites, and many other decisions.

The EPA may strengthen its policies, but that is no guarantee that states and localities will do so. And that is where it matters most for sustainability. The alienation of land use planning from environmental planning in the United States also alienates the EPA from on-site decision making in many instances. The EPA itself is an agency made up of more than 18 mission statements from Clean Air and Clean Water, to Pesticide regulations and environmental impact statements. The EPA has different roles in these decisions, but it is almost always the convening stakeholder. The EPA can be a partner with national projects and community groups, who can develop recommendations to EPA, and sometimes binding agreements. As a decision maker, the EPA exchanges large amounts of important environmental information and develops recommendations. As a capacity builder, the EPA exchanges information, develops recommendations, and can develop binding agreements.

In most public participation processes in the United States, the government is the convening stakeholder. Most public participation consists of meetings and information exchanges, with a very small part of any one potentially impacted population getting notice or actually getting involved. Some communities are so concerned about the public health effects of accumulating uncontrolled and unmeasured environmental impacts they monitor the environment themselves. Their "notice" is their own observations, scientific and otherwise, of changes in the environment. Because of this groundswell of community concern and greater knowledge of what public participation opportunities exist, the range of public participation tools that meet minimum legal requirements has expanded. The growing range of public participation tools is a good development for sustainability decision making. These tools will need to be embraced as in the model plans of land use at state and tribal levels.

Reference

Zazueta, Aaron Eduardo. 1998. *Policy Hits the Ground: Participation and Equity in Environmental Policy Making.* Washington, DC: World Resources Institute.

SINGLE INTEREST CONSULTATIONS: GOOD OR BAD FOR SUSTAINABILITY?

Another generally one-time, information-focused, and informal form of participation are single interest consultations, usually industry consulting with the permitting environmental agency or local land use authority. Industry may need a permit from the government and consults with the permit-issuing authority, usually a state environmental agency with authority to control permits delegated to it by the EPA. The potential industry permit holder may encourage parts of the community with employment promises, especially those with values of economic development. Generally, the industry permit applicant is seeking to comply with the environmental laws and regulations in the least expensive manner. Most industrially based environmental information is self-reported, and most industries must emit a certain amount before any environmental regulation occurs. When the permit is issued, it does not necessarily limit the environmental impact, but rather sets some limits and requires industry to continue to self-report its emissions and pollution to the EPA or state environmental agency. The number of employees a state environmental agency uses to help industry write permits is often contrasted with the number of people it employs to enforce environmental laws. Some states argue that good economic development requires agencies to offer as much compliance assistance as possible to all industries, especially if they want to comply with environmental laws and attract economic development at the same time. Federal and state environmental agencies often have "compliance centers" for big and small industries.

Others have criticized these meetings as closed deals with industry. This is also aimed at local government land use decision making. Environmentalists and communities want to be part of these single-interest consultations. They want all the information to be transparent and accessible. Industry and government argue that if this were the case, it would not be a single-interest consultation. Some state and federal environmental agencies are allowing other stakeholder groups to meet with them as a single interest, but usually long after the permit application has been submitted and generally only in controversial cases. In terms of sustainability, the information vacuum around environmental and human health impacts in industry and government single consultations runs counter to the free flow of information necessary for inclusionary and deliberative dialogues.

WORKSHOPS AND FORUMS

Workshops and forums are generally weak in terms of direct decisional power, but they do empower and engage stakeholders and increase the flow of information. Community science workshops and labs are run by the French government to increase the capacity of communities to engage in these decisions. Traditionally, they require some face-to-face contact with some representative of industry or government. With rapidly advancing communication technology, it may be able to decrease in-person transactions. In terms of sustainable decision-making regimes, this would be necessary. Current workshop and forums tend to be scattered and characterized by poor outreach results. They offer a tremendous and tried ability to increase the capacity of the community to engage in these decisions. Unfortunately, if communities are given absolute power, they may come to conclusions that challenge both private property and environmental norms. A community could decide to damage all natural resources irreversibly to make a huge, short-term profit. More likely, a community could decide, after expensive and extensive scientific study, that the environmental and/or public health risks of a given project exceed the benefits. Industry would resist this type of process because it is financially riskier for them to invest in a long-term, community-contingent decision-making process.

ROUNDTABLES AND INFORMAL POLICY DIALOGUES

Roundtables imply inclusion and equality in their design. Everyone is facing everyone else. Information is exchanged, it is applied to environmental issues, but not much is decided. It is a deliberative process in that the perspectives of the other stakeholders are considered, but it is not truly deliberative in the sense necessary for sustainability because the common good is not the goal of every stakeholder. The developer wants her land development permit approved quickly, and wants it to allow her the most profitable use of the land. That use may be a manufacturing plant with large environmental impacts and potential employment. The company wants its environmental permits approved as quickly as possible so it can operate in the most profitable manner. Both local and state elected officials know that most U.S. constituencies currently value economic over environmental security. If they are accused of causing job loss through overly restrictive environmental policies, their chances of reelection decrease. They are highly motivated to make all the permit processes as fast as possible. Deliberative dialogues are not fast, and if the capacity building of the community takes too long or requires too much environmental disclosure or cleanup by industry, then they may withdraw their proposal for the land use permit and for any environmental permits.

As in most public participation processes, emphasizing how a decision could personally affect individuals increases community engagement. Sustainability would require that this process not be a piecemeal consideration, as it today. Today, the permit is considered and usually approved without knowledge of past, present, or future current environment impacts. Because most environmental information from industry is self-reported, many do not know when they have crossed the threshold of emissions and when it is necessary to get a permit. Sustainability will require far more accurate and complete environmental information. It will require knowledge of current accumulated impacts on the ecology, as well as history of the land. Complete emissions from industrial, commercial, and residential sources all add to the cumulative environmental impact, or footprint, of human habitation. To the extent roundtables and informal policy dialogues contribute to deliberation and inclusion, they are vehicles for sustainable decision making.

CONSENSUS AND NEGOTIATION: CAN THEY BE SUSTAINABLE AND FAIR?

Next there is a series of formal meetings that are multistakeholder and consensus based. They are usually held at the federal level, often at Washington, DC headquarters. Most of these are federal advisory committees covered by the Federal Advisory Committee Act. All federal agencies have advisory committees governed by the Federal Advisory Committees Act. This federal advisory committee is called the National Environmental Justice Advisory Council, or NEJAC.

Usually the next step in the process of developing a public participation plan is to match the issues of the decision to stakeholders. This is a difficult process. Many environmental justice communities insist they speak for themselves, and most other communities in developed nations assume they can. Most of the time, U.S. communities do not participate meaningfully in the environmental decisions that affect them. This step of the current model of public participation planning is usually narrowly construed by the convening stakeholder, the government. In each broad, general stakeholder group of industry, community, and government, there are internal stakeholders. Big industries may have different interests than small industries. Government has layers of federal, regional, state, local, and research intergovernmental relations. In most large environmental decisions, the federal environmental or state environmental agency is the convening stakeholder.

International Public Participation: The Aarhus Convention

The Aarhus Convention resulted from a meeting in Aarhus, Denmark, in 1998. The UN ECE Convention on Access to Information, Public Participation in Decision-making and Access to Justice in Environmental Matters was adopted on June 25, 1998, in the Danish city of Aarhus at the Fourth Ministerial Conference in the Environment for Europe process. Parties have subsequently met in Lucca, Italy (2002) and Almaty, Kazakhstan (2005). In May 2003, parties met in Kiev, Ukraine in an extraordinary meeting to develop a pollutant transfer and registry protocol. Parties are scheduled to meet next in Riga, Latvia. In all, 35 countries and the European Union are signatories. Parties are mostly European and the United States is not a party. Aarhus is a project of the United Nations Economic Committee for Europe and the UN Environment Programme.

The Aarhus Convention emphasizes public involvement in environmental decision making and government accountability. As such, the convention also focuses on providing citizen access to environmental information. The convention essentially requires signatories to establish environmental "Freedom of Information" laws. The convention requires signatories to implement a legal process that allows citizens redress when environmental information covered by the agreement is not readily provided. These goals are said to be represented by three pillars used to describe the convention:

- Access to Information

- Public Participation

- Access to Justice

The Aarhus Convention is a new kind of environmental agreement. It links environmental rights and human rights. It acknowledges that we owe an obligation to future generations. It establishes that sustainable development can be achieved only through the involvement of all stakeholders.

The convention links government accountability and environmental protection. It focuses on interactions between the public and public authorities in a democratic context. Finally, it forges a new process for public participation in the negotiation and implementation of international agreements.

The convention is not only an environmental agreement; it is also a convention about government accountability, transparency, and responsiveness. The Aarhus Convention grants the public rights and imposes on parties and public authorities' obligations regarding access to information and public participation and access to justice.

The Aarhus Convention is a new kind of environmental agreement. It links environmental rights and human rights. It acknowledges that we owe an obligation to future generations. It establishes that sustainable development can be achieved only through the involvement of all stakeholders. It links government accountability and environmental protection. It focuses on interactions between the public and public authorities in a democratic context and it is forging a new process for public participation in the negotiation and implementation of international agreements.

United Nations Environment Programme Website Summary

The Aarhus Convention evolved from the Rio Declaration and ECE concerns for sustainability. Although regional in scope, the significance of the Aarhus Convention is global. It is by far the most impressive elaboration of principle 10 of the Rio Declaration, which stresses the need for citizens' participation in environmental issues and for access to information in the area of environmental democracy so far undertaken under the auspices of the United Nations.

References

Aarhus Convention Implementation Guide at: www.unece.org/env/pp/acig.htm.

Hayword, Tim. 2005. *Constitutional Environmental Rights.* London, UK: Oxford University Press.

United Nations. 2006. *Your Right to a Healthy Environment: A Simplified Guide to the Aaurhus Convention on Access to Information, Public Participation, in Decision Making.* New York: United Nation Publishing.

Collaborative Approaches

Good Neighbor Agreements

A good neighbor agreement is an enforceable contract or a specific agreement that details a set of commitments that a corporation is required to make in order to demonstrate its accountability to the community. These commitments address the needs identified by the affected employees and other community residents. They often go beyond the requirements of local land use rules, environmental laws, or business laws. They often focus on the environmental and public health impacts of the industry on the community and sometimes on the workers. Some typical commitments are to study and reduce toxic chemical usage and hazardous waste production, to provide funds for an independent expert for community to review the industry's activities, or to conduct periodic plant inspections, which may be announced or unannounced.

Many communities want sustainable development but do not think that current environmental laws and land use regulations give them enough control of an industrial facility. The exact provisions of a given good neighbor agreement are based on the type of industrial facility and the concerns of the neighborhood. There is often a focus on sustainable development and long-term good relations, as in neighbors.

COMMUNITY BENEFITS AGREEMENTS

A community benefits agreement occurs in the context of economic development in a given community. It is a contract a real estate developer signs with organized representatives of the community specifying the range of community benefits they will give the community as part of the development. It is generally preceded by a negotiation. Community groups agree to support the development at the stages of government approval necessary for the development to go forward. It is possible for many developments to go forward without community approval but with agreements with the city or municipality. Cities actively pursue economic development and may make agreements that reduce land use and environmental rules in one part of the city but do not cover where the benefits of the development occur, if at all. Once the development is built, there is typically a provision for continued monitoring of the development and level of promised benefits. Benefits often take the form of jobs, decrease in environmentally degrading activities, or the provision of desired community amenities.

Reference

Community Benefit Agreements, www.goodjobsfirst.org/pdf/cba2005final.pdf.

Good Neighbor Agreements, www.cpn.org/topics/environment/goodneighbor.html.

Gross, Julian. 2005. *Community Benefits Agreements: Making Development Projects Accountable.* San Francisco: Good Jobs First.

SUSTAINABLE DEVELOPMENT

Sustainable development links the goals of traditional development—economic improvement and social betterment—to environmental protection. The most frequently quoted definition comes from a report for the World Commission on Environment and Development called, "Our Common Future," written in 1987 by Gro Harlem Brundtland, then the prime minister of Norway. She wrote that sustainable development is "development that meets the needs of the present without compromising the ability of future generations to meet their own needs."

Sustainable development assumes continuous economic growth, without irreparably or irreversibly damaging the environment. Human population growth is difficult for this model because it is difficult to place an economic value on the lives that exist in the future. Some environmentalists challenge the assumption of growth at all. The fundamental battleground for this emerging controversy is one of values. The continued prioritization of economic growth over environmental protection, combined with population increases, may have irreparable impacts on the environment; therefore any sustainable development policy would require governments to place constraints on development that have not been present before.

Countries committed to development to meet the basic needs of their contemporary populations face difficult choices for sustainability, including the choice of which energy resources to use. If these countries choose to develop a fossil-fuel–dependent economy, the additional contribution to worldwide ecological crises like climate change will certainly result in more ecological damage. Alternative fuels and technologies, however, are less readily available and often much more expensive to procure. To develop without reliance on fossil fuels, these countries need access to technologies and funding for investments that offsets the costs of those options.

Race and Gender Differences in Environmental Risk Perception: A Big Divide

An important part of increased public participation is how individuals actually perceive environmental risks, as well as how the environmental risks impact the ecosystem and physical aspects of public health. The perception of personal risk from belching, unfiltered petrochemical smokestacks operating all the time has caused increased stress leading to hypertension and increased risk of stroke and heart attack. The scientific causality of the risk may be certain, but the impact remains controversial if the perception of it is harmful. The range of differences in perception of environmental risk is astounding. Enormous differences in perception of risk exist between residents, as well as between professional risk assessors. Gender, race, age, worldview, income, education, perceived control over exposure to health risks, employment status, and political attitudes all shape the perception of risk. To the extent these factors are part of the inclusionary new dialogues of sustainability, they will shape some of the first policies of sustainability.

Researchers have analyzed the risk perceptions of different demographic groups and some countries. What often stands out for purposes of equitable considerations in sustainability is that the most privileged groups in a society are often the ones who underestimate risk. Privileged groups may not self-recognize or acknowledge their own privilege. Generally, research in the United States indicates the following characteristics of environmental risk perception:

- Females see more risk from nuclear power, nuclear waste, and nuclear weapons.

- Nonwhite populations perceive their home as exposed to environmental health problems far more than white populations.

- Nonwhites and females perceives risks about one-third higher than white males in most risk categories

- The worldview or perspective of professional risk assessors influences their decisions.

These generalizations are dynamic and may change over time, but they are important for sustainability because including populations that have been overexposed to environmental degradation will confront those with privilege and power. Unacknowledged privilege becomes less controversial when better environmental and ecological monitoring begins to establish baselines that show which areas are environmentally burdened and which are environmentally benefited.

Van Jones

Van Jones graduated from Yale Law School in 1993 with the goal of making a difference. He has. Jones is a civil rights attorney and works in social justice issues, but that is not all. He is the founder and president of the Ella Baker Center for Human Rights and Green for All, two nonprofit organizations based in California. The Ella Baker Center works to keep youth out of prison and prevent youth violence. Green for All helps build an inclusive green economy and focuses on creating "green pathways out of poverty" and tirelessly fights for expanding the coalition against global warming. Jones plans on incorporating the goals of the Ella Baker Center and Green for All by creating a program to ensure that low-income and minority youth have access to the coming wave of "green-collar" jobs.

Jones believes that environmentalists are trying to solve the global environmental problem alone, instead of incorporating all sectors of society. He argues that involving the majority of the people by explaining the opportunities available with green technology will result in a much more effective campaign to reduce global warming and local environmental degradation. He gives a good example: If a family knows that if the air gets better where they live, their child will not get asthma, which will save them per child, $10,000 a year in hospital visits, health care costs, and missed work. Knowing this, they will support the closing of a plant that emits the pollution.

References

Van Jones Home and Bio at www.vanjones.net/page. php?pageid=3,andcontentid=29, 6/8.

Roberts, David. A Van With a Plan—An Interview with Van Jones, Advocate for Social Justice and Shared Green Prosperity, Mar 20, 2007. www. grist.org/news/maindish/2007/03/20/vanjo nes/, 7/16/2008.

URBAN LAND: GROWTH OF URBANIZATION AROUND THE GLOBE

Urbanization most generally refers to the redistribution of populations from rural environments to urban settlements. The impact of this movement is usually city centers that are densely populated compared with mostly sparsely populated rural areas. Cities are often places where money and wealth are centralized, and modern business usually has its epicenter in urban locations. This desire to increase one's livelihood is often the impetus for urbanization. Urbanization is particularly appealing as small rural farms are often supplanted by large agribusiness; it is becoming increasingly difficult to maintain a sustainable living on a small farm. Urbanization has increased over the last 100 years and has taken a number of different forms including suburbanization that resulted in an outgrowth from city centers to a more sprawling car-centric design. There is some evidence that suburbanization is facing a decline.

Urbanization has most often been a process in which an increasing percentage of the total population of a region lives in cities or suburbs of cities. Urbanization has usually been closely connected to industrialization, specifically the advent of factory production at the turn of the 21st century. Before the global trend of urbanization, the human population had lived a rural or agrarian lifestyle. In 1950, less than 30 percent of the world's population lived in cities; by 2000 that number was almost 50 percent. It is expected that by 2025, 65 percent of the world's population will live in cities. Urbanization is growing exponentially; the result is overcrowding of cities throughout the world.

Currently developed nations have a higher percentage of urban residents than less developed countries. This will not always be the case; there is evidence that the largest urban growth will occur in less developed countries. The growth of urban areas is often the result of increased populations, especially in underdeveloped and developing countries. Migration to urban areas is also a factor in urban growth. Increased urbanization is usually the result of internal migration within a nation from more rural areas to rapidly growing cities. Cross-national migration, however, is also a contributing factor in urbanization patterns.

Urbanization may have a lower ecological footprint than suburban, sprawl development. The increased density leads to many decreased environmental impacts from transportation, building materials, and energy. The need to understand and reclaim urban environments is very high under sustainable development. The Urban Atlas Portal is an international collaboration between cities to develop ways and tools to understand the ecological capacities necessary to provide access to and sustain ecosystem services in urban areas. It developed an atlas that can be navigated via the Web, launched December 2009 (www.stockholm resilience.org). The atlas will show ecological and social data, maps, charts, three-dimensional images, and rough model building. This system makes primary environmental data public and allows stakeholders to share information with each other across regions. Collaborations between urban areas are greatly increasing and will provide the basis for better data about equity and sustainability.

See also **Volume 3, Chapter 1: The Urban Context of Equity in the United States; Volume 2, Chapter 4: Urban Sprawl.**

Brownfields, Greenfields, Greyfields

Brownfield land or brownfields are abandoned plots of land that were once used for commercial or industrial facilities. This past use often resulted in low concentrations of contamination by hazardous waste or pollutants. Brownfields are specific plots of land that have a potential to be reusable once the land has been sufficiently cleaned up. Most brownfield sites have been left unused for significant periods; however, as land in certain locations has become more scarce or expensive, brownfield sites are worth enough to bring up to safe standards and redevelop. Furthermore, with increased precision and new techniques, the ability to bring brownfield land up to safe standards has become scientifically and economically feasible.

Brownfield redevelopment is still not perfect, and some projects are abandoned because of rising costs caused by unknown contaminants that exceed the initial evaluation. Most Brownfield cleanup projects are for commercial use; however, there are some projects underway to determine if brownfields can be used to grow crops. The intent is twofold: first, to help with the cleanup process of the soil and second, to contribute a more efficient production of biofuels. The regulation and cleanup of brownfield land is regulated by the EPA. The EPA works with individual states based on state codes to provide technical assistance and cleanup of brownfields, as well as to determine sources of funding to ensure that brownfields are given new life.

Greenfield land is a piece of undeveloped land that is either being used for agriculture or is currently undeveloped. Greenfield land is sometimes the result of decommissioned industrial sites. These plots of land often need time to return to the condition they were in before construction of a given industrial plant. If a former industrial site is decommissioned, it cannot be reused until the land has had a chance to heal. Not all greenfield sites consist of land that is in the process of healing. Some greenfields are greenbelts that have prohibitions against development. Greenbelts are designed to protect the unique character of undeveloped land within areas of extensive development.

Greyfield land is land that was once thought to be economically profitable but eventually became obsolete and outdated. The term is usually applied to areas that were once considered viable retail and commercial plots of land that suffered from lack of reinvestment. Greyfields are usually not contaminated; instead they are usually abandoned because of larger developments nearby. Greyfields may have a dormant value because they are often equipped with an underlying infrastructure, such as plumbing and sewage systems, that may be used if the land was redeveloped.

References

De Sousa, Christopher. 2008. *Brownfields Redevelopment and the Quest for Sustainability.* Oxford, UK: Elsevier Science.

Dixon, Tom et al., eds. 2007. *Sustainable Brownfields Regeneration: Livable Places from Problem Spaces.* Hoboken, NJ: Wiley-Blackwell.

Smart Growth

Smart growth is a government-supported community concept designed to cover a range of developmental and conservation strategies that help protect our environment and make our communities more attractive, economically stronger, and more socially diverse. Smart growth community development includes newly built housing projects, as well as changes to existing developments. Smart growth also has a component of policy considerations to facilitate long-term goals and changes to communities to foster the goals of environmentally conscious and sustainable living. Smart growth principles include mixed land uses and taking advantage of compact building design to preserve space and create land use efficiency. There is also a strong community component to the smart growth concept that includes creating walkable neighborhoods and fostering distinctive, attractive communities with a strong sense of place. The idea is not only to preserve land but also to create a sense of community involvement in making good use of the limited available land space. Furthermore, there is an element of extending outward from the smart growth community and creating a variety of transportation choices and making development decisions predictable, fair, and cost effective.

The EPA monitors and prescribes the recommendations for smart growth communities. By setting standards and goals, the EPA encourages small communities to take on the smart growth goals and extend them outward to include changes in water use, transportation, and air quality, as well as creating new uses of available land and preserving open spaces, farmland, and natural beauty. Neighborhoods that engage in smart growth planning not only benefit from the underlying goals of the project but also set examples for other communities by demonstrating that smart growth plans are both and increase the standard of living for inhabitants. Smart growth planning considers a variety of goals including climate protection, environmental protection, and public health. The goals of smart growth communities are based on the idea that people can reinvent the neighborhoods and cities they live in to comport with fairer housing standards and long-term sustainability goals.

The smart growth planning model also encourages a greater worldview. It encourages mixed-use development and housing choices to encourage diverse communities. The planning also fosters walking and biking as feasible alternative transportation. These ideas encourage equitable living circumstances for a wider range of individuals. The public health aspect of smart growth is especially impacted by transportation alternatives by creating goals that decrease car pollution in urban areas. Through the diversification of communities, these goals can affect a diverse group of individuals. The smart growth model encourages individuals and communities to take an active role in shaping their environment. People can participate by riding their bike or reusing a space for sustainable development.

References

Davenport, John, and Julia L. Davenport. 2006. *The Ecology of Transportation: Managing Mobility for the Environment.* New York: Springer.

Goldfield, David. 2006. *Encyclopedia of American Urban History.* Thousand Oaks, CA: Sage.

Harris, Leslie M. 2002. *In the Shadow of Slavery: African Americans in New York City 1626–1863.* Chicago: Chicago University Press.

Kusmer, Kenneth L., and Joe W. Trotter. 2009. *African American Urban History since World War Two.* Chicago: Chicago University Press.

Porter, Douglas. 2002. *Making Smart Growth Work.* Washington, DC: Urban Land Institute.

Tolley, Rodney. 2003. *Sustainable Transport: Planning for Walking and Cycling in Urban Environments.* Cambridge, UK: Woodhead Publishing.

CHAPTER 3

Government and United Nations Involvement

ROLE OF GOVERNMENT IN ENSURING EQUITY UNDER SUSTAINABILITY

Education: Environmental, Multicultural Education

There are many challenges to sustainable decision making. One challenge is how to be fair in the process. The stakes become higher than in usual environmental decision making because of people's personal health, irreversible environmental effects, and loss of economic value as a proxy for social good. Our current state of environmental knowledge is characterized by ignorance. No one yet knows how to make a dense, old urban area ecologically sustainable. The stage is set for a large, institutional restructuring of education.

Environmental education in the United States has historically been characterized as liberal, or extracurricular, or not credible. Poor schools and communities without schools are often without the luxury of environmental education. At the college level, environmental education does not fit neatly into anyone discipline. In Australia the civil engineering curriculum dominated environmental education; the philosophy department covered it in New Zealand. Education reflects other institutional arrangements in society, with values acting in support of these arrangements. In the United States, science tends to support those areas that produce profit and protect profit and economic development. Because science is the basis of U.S. environmental regulation, one of the values of U.S. environmental regulation is profit. This conflicts with newer values of inclusion, precaution, and general fairness, all of which will be expensive in economic terms. They may cost so much that present generations will not be able to recoup their investment in these processes in their own lifetimes. One social institution that leverages all other values is education.

In all institutional and private interactions with nature, we are environmentally educated. Narrow concepts of environment, which exclude the built environment, things that humans cannot perceive, or dynamics that cannot be proved to cause anything yet observably exist, tend to be a perspective of more affluent nations. Broader concepts of environment include ancestors, dental care, and future generations. By reconceptualizing and prioritizing environmental education in all its structure and presence, a better understanding of environment develops in a way that facilitates sustainable decision making.

The focus of environmental education is the preparation of all members of society for new processes of inclusionary environmental decision making that makes up part of equitable aspects of sustainability. The range of environmental education could expand in many ways. It could include a working knowledge of world ecosystems as well as one's own. Right now, in most nations it requires increasing the capacity of new participants in the analysis of environmental issues for effective sustainable policymaking and management. Environmental education will also include the methods of public participation and environmental planning. Issues like citizen monitoring, cumulative risk assessments, and ecological risk assessments will enter environmental studies curricula as public policies refine their applications. Environmental education will include the study of currently unknown technologies.

Almost all education is seen through one or more cultural lenses. In the United States, the range in the perception of risk from environmental hazards from female to male and from white to nonwhite is large. The inclusionary dialogues around the ecology and public health of a community will span these gulfs and others. Communication around concepts of sustainability is necessary to span them. Issues of sustainability need to be taught in a multicultural context with many different approaches to "environment."

The perception of environmental risk varies greatly by gender, race, income, and education. Highly educated white males have the lowest perception of risk; all females and all nonwhite people share about the same perception of risk. Urban and suburban, rural and nonrural perceptions of environment vary all around the world. What is sacred to one group could be a toxic threat to another. White males have been shown to have an aura of invincibility about environmental risk perception, which some argue is a measure of unacknowledged environmental privilege. In the United States, the greater the education and income of the white male, the more likely he was to agree with putting communities at risk without disclosure. Many of the older, U.S. risk models were based on a 150-pound white male, around age 30. This is one of the healthiest and least vulnerable segments of U.S. society.

Risk perception and risk models can be historically skewed to one demographic that is politically and economically powerful. Environmental educators will have to engage these topics because of their teaching environments. World populations are becoming more urban and more multicultural. Experiential learning experiences in the environment of urban areas reveal important environmental baseline information necessary for application of sustainable decision-making principles like the precautionary principle. They also expose areas of toxic waste accumulation, environmental injustice, and irreversibly damaged natural systems.

Multicultural environmental education is necessary for sustainable decision making because shared and collective perceptions of the environmental issue facilitates sustainable decision making. Knowledge about past environmental acts and the increased scale of human impact on the environment now make the consequences of environmental actions from the past occur within a lifetime.

References

Allen-Gil, Susan et al. 2008. *Addressing Global Environmental Security through Innovative Educational Curricula.* New York: Springer.

Josephson, Paul. 2004. *Resources under Regimes: Technology, Environment, and the State.* Cambridge, MA: Harvard University Press.

Leal Filho, Walter. 2006. *Sustainability in the Australasian University Context.* New York: Peter Lang.

Leal Filho, Walter, and Mario Salomone, eds. 2006. *Innovative Approaches to Education for Sustainable Development.* New York: Peter Lang.

Orr, David. 1994. *Earth in Mind: On Education, Environment and the Human Prospect.* Washington, DC: Island Press.

Public Participation in Decision Making

Governments on all levels have the power to call interested persons together in one forum. The power to convene a discussion can significantly affect the path of development and change. Governments can exercise this power along a broad spectrum of activities. In the past, this power has often been used to communicate policy decisions that already have been made with many of the interested parties absent or ignorant that such policies were being made. When decisions are reached without the knowledge or consent of affected people, those policies are much less likely to become self-enforcing. The hallmark of democratic decision making is the idea that the governed have given their consent to the government and will therefore support its decisions. Sustainability in environmental decision making requires governments to use their power to convene interested parties in a different way. Instead of convening people to communicate after an important decision has been reached, sustainable decision making requires government to use this power to convene proactively to formulate environmental policy with the active, informed participation of the people who will ultimately live with the consequences of that policy.

This role represents a change for many bureaucrats, and they may be uncomfortable with the combined roles of educators, facilitators, communicators, and decision makers. Understanding the forms of public involvement, the assistance that is available, and how to deploy a multi-stakeholder process fairly and effectively is the path of the future.

References

Kasemir, Bernd, and Jill Jager. 2003. *Public Participation in Sustainability Science: A Handbook.* Cambridge, UK: Cambridge University Press.

Tilbury, Daniella, and David Wortman. 2004. *Engaging People in Sustainability.* Gland, Switzerland: International Union for Conservation of Nature and Natural Resources.

Barriers

PRIVILEGES AND RIGHTS

Many of the participants in environmental and economic decision making are not used to sharing access to power, or sharing the right to make decisions. This power to exclude others comes from a history of land ownership and includes the development of modern concepts of private property.

PRIVATE PROPERTY

Private property is one of the strongest values of industrial and even postindustrial societies. Private property can represent many things—status, power, security, law and order, and liberty and freedom. It allows a person to own, or control, land. "Persons" have been legally expanded by the courts to include corporations, ships, fetuses, and other entities other than residential average citizens.

The big concern of the writers of the U.S. Constitution was the control by government of individual liberty and freedom, especially as expressed in land. The takings clause of the Fifth Amendment to the U.S. Constitution states: "nor shall private property be taken for public use, without just compensation." Most governments can take, or nationalize, property for the state. They do not necessarily need a public purpose, nor do they need to pay for it. In the United States, the government can take private property, but it must be for a public purpose and the government must pay a fair price for it. This is called the eminent domain of the state to control the land. The U.S. government is one of the largest landowners in the United States and has this constitutional power. Some sustainability advocates view this as a hopeful sign for future land use planning for sustainability. Currently, most U.S. local governments seek to avoid using the taking power because of the political controversies that can occur. However, there is increasing pressure from environmentalists, international groups, community groups, and other voters to stop environmentally unsustainable uses of private property. This is a thorny issue. Can the liberty and freedom, status, power, law and order, and security functions of private property be retained in a land-use system based on ecological sustainability?

References

Freyfogld, Eric T. 1993. *Justice and the Earth: Images for Our Planetary Survival.* New York: The Free Press.

Freyfogle, Eric T. 2003.*The Land We Share: Private Property and the Common Good.* Washington, DC: Island Press.

Fuchs, D. A. 2003. *An Institutional Basis for Environmental Stewardship: The Structure and Quality of Property Rights.* New York: Springer.

Weak Communication between Local and State Government: An Environmental Problem

Local land use practices and policies do not communicate well with state environmental agencies. This is a very weak link in the U.S. environmental decision-making process. State environmental agencies issue permits to industry to emit chemicals into the land, air, and water. Societies with a high value on economic development through industrialization will place a priority on the speed of the transaction because time can erode profit. Industry is always complaining about the time it takes the government to process permit applications, renewals, and modifications while at the same time resisting even self-reporting of some environmental information. Local government land use decisions regarding industry is often in the same dynamic when industrial economic development is sought.

Local land use policies and state environmental policies conveniently do not share much information about prospective new development. Because speed of transaction is important for economic development, time consuming, and troublesome, stakeholders that could stop a project are not included. Or, if the law requires some type of notice to the community, citizens are included in ways that are not meaningful. Industry will face a big challenge from the equity component of sustainability because this type of decision making is not inclusive and does not accurately represent past, present, or future environmental impacts.

When developing a public participation plan, leaders must plan for capacity building of some stakeholders, even if it is time consuming. The environmental dimension of sustainability will greatly expand the period of decisions because so little is currently known about measuring, monitoring, and restoring human habitation to an ecological balance. The traditional, narrow approach of simply matching issues with stakeholders offers a framework useful for equitable decision making under sustainability, but with greatly expanded time requirements. It is likely that this will be an expensive cost and will increase the risk that a project is denied or developed with such costly mitigation requirements that no profit is possible.

UNITED NATIONS

The United Nations is not a government in the strict sense; it is an association of governments. Together the member states explore ways in which they may act collectively on issues affecting global peace and security. Sometimes, the way the UN functions is to assist member nations in articulating norms of behavior. When these norms are not written into enforceable treaties, this kind of activity is called "soft law." Soft law can be operated on voluntarily by individual groups and by nongovernmental organizations. "Hard law" is an international norm of behavior backed by enforceable treaties. Sometimes soft law becomes hard law over time, as international norms first articulated in soft law come to be expected and relied upon.

The UN works in a number of ways to promote and develop soft and hard law around the issues of sustainable development. It sponsors scientific research and publishes information critical to policy development. It also sponsors many conferences at which soft law principles of sustainability emerge and consensus develops. It promotes the development of nongovernmental organizations with goals compatible with these principles. In addition, it offers member governments consulting services, work plans, and training.

References

Andersen, Steven O. et al. 2002. *Protecting the Ozone Layer: The United Nations History.* London, UK: Earthscan.

Murphy, Craig N. 2007. *The United Nations Development Programme: A History.* Cambridge, UK: Cambridge University Press.

Weiss, Thomas G. 2005. *UN Voices: The Struggle for Development and Social Justice.* Bloomington: Indiana University Press.

United Nations Environmental Programme (UNEP)

UNEP coordinates all the environmental activities of the UN. It began at the United Nations Conference on the Human Environment in Stockholm in 1972. Many scientists and environmental activists were concerned that many environmental problems spanned the boundaries of nations and that rising world populations could make these problems worse in ways no one nation could effectively handle. After the conference the United Nations passed Resolution 2997 in December 1972, establishing the United Nations Environment Programme as a permanent institution charged with protection and improvement of the environment. Resolution 2997 also charges the UNEP with the missions of promoting international cooperation on the environment; reviewing global environmental issues so that governments give them adequate consideration; promoting the acquisition, assessment, and exchange of environmental knowledge; and reviewing the environmental impact of environmental policies on developing countries.

The UNEP is governed by a Governing Council of 58 members elected by the UN General Assembly for 3-year terms. Member seats are allocated on a global regional basis. The headquarters is in Nairobi, Kenya with six regional offices around the world. It has seven divisions: Early Warning and Assessment; Environmental Policy Implementation; Technology, Industry and Economics; Regional Cooperation;

Environmental Law and Conventions; Global Environmental Facility Coordination; and Communication and Public Information. UNEP is a big player in major international environmental initiatives. They publish many books and reports. UNEP's medium-term strategy for 2010–2013 is to prioritize climate change, disasters and conflicts, ecosystem management, environmental governance, harmful substances and hazardous wastes, and resource efficiency-sustainable consumption and production. For more information on UNEP, see www.unep.org.

Reference

Andersen, Steven O. et al. 2002. *Protecting the Ozone Layer: The United Nations History.* London, UK: Earthscan.

Civil Society Programs

The United Nations Development Program (UNDP) works with civil societies at all levels to promote the Millennium Development Goals (MDGs). UNDP considers civil societies essential for its goals of national ownership, accountability, good governance, decentralization, democratization of development cooperation, and the quality and relevance of official development programmes. It uses a Civil Society Advisory Committee to facilitate communication between civil societies and UNDP.

UNDP works with civil societies at the local, regional, and global levels in six thematic areas: Democratic Governance, Poverty Reduction, Crisis Prevention and Recovery, HIV/AIDS, Energy and Environment, and Women's Empowerment. Supporting the capacity development of civil society is central to this partnership. UNDP gets baseline information on a given civil society to provide an accurate assessment of its functional ability so it can later determine if it is useful. Examples of civil societies that work with the UN in this manner are nongovernment organizations (NGOs), cooperatives, trade unions, service organizations, community-based organizations, indigenous peoples' organizations, youth and women's organizations, academic institutions, policy and research networks, and faith-based organizations. Through the UNDP, the UN works with 160 countries, and these offices are usually the first contact of civil societies. Engagement by the UN with civil societies can be a delicate political issue and is usually determined by the UNDP and *Civil Society Organizations: A Policy of Engagement.*

NONGOVERNMENTAL ORGANIZATIONS

Governments are increasingly working with NGOs and business to achieve sustainability goals. The UN works to make NGOs part of dialogue on important global issues. The UN Economic and Social Counsel is particularly involved. It holds meetings and hearings with NGOs through the commissions on Sustainable Development, the Status of Women, and Population and Development. The MDG program also seeks to involve NGOs. The UN also hosts a regular meeting of NGOs.

References

Betsill, Michele M., and Felix Corell, eds. 2007. *NGO Diplomacy: The Influence of Non-governmental Organizations in International Environmental Negotiations.* Cambridge, MA: MIT Press.

Major Conventions

THE GLOBAL COMPACT

The Global Compact is an agreement of businesses to support human rights, labor standards, anticorruption, and environmental responsibility. Each participating business agrees to seek to incorporate these goals into their business practices. Although businesses are not considered part of civil society by all analysts, the Global Compact is a unique example of a UN agreement among nongovernmental entities.

THE MILLENNIUM DEVELOPMENT GOALS

In 2001, all 189 United Nations member states adopted the United Nations Millennium Declaration. This declaration laid out eight goals for human society and government for the this millennium:

Goal 1: Eradicate extreme poverty and hunger

Goal 2: Achieve universal primary education

Goal 3: Promote gender equality and empower women

Goal 4: Reduce child mortality

Goal 5: Improve maternal health

Goal 6: Combat HIV/AIDS, malaria and other diseases

Goal 7: Ensure environmental sustainability

Goal 8: Develop a global partnership for development

The 8 goals are further divided into 21 targets within the framework of the MGDs. These goals and their targets are considered the leading goals and indicators of sustainable development at the present time.

The benchmarks of the MGDs are rooted in international development and are to be carried out by a number of UN member states, as well as international organizations. The MDGs were officially established at the Millennium Summit in 2000 in New York City. At the time, this was the largest gathering of world leaders (eventually surpassed by the World Summit in 2005). The MDGs are derived from the United Nations Millennium Declaration that was adopted at the Millennium Summit.

The MDGs wanted to prompt progress and set a goal to achieve all eight of the goals by 2015. Thus far progress has been exceptional by some nations and completely lacking by others. The Multilateral Debt Relief initiative to effectuate debt cancellation for heavily indebted poor countries was in part to help realize some of the MGDs. Although NGOs are not officially part of the MDG agreement, they play an important role in pursuing and achieving the MDG.

PROGRAMS AND OFFICE

The following UN projects seek to help civil society achieve sustainability goals:

- Nongovernmental Liaison Service
- U.N. Civil Society Web site (allows lay user to learn about and navigate UN resources)
- United Nations Development Programme
- Economic and Social Counsel
- Community Commons
- Department of Public Information, NGO Section
- Millennium Development Goals Campaign Office

References
UN Global Compact, www.unglobalcompact.org/index.html.
UN General Civil Society Hearings, www.un.org/ga/civilsocietyhearings/.
UN Non-governmental Liaison Service, www.un-ngls.org/.

The Role of Women

The role of women in sustainable development is very important. Women are often closest to the environment, always have direct involvement in population control, and can be politically oppressed. The role of women in any given culture can be heavily influenced by religion, education, and economic and social institutions. These social institutions also affect the values of a given nation, as well as the capacity of that nation to engage in sustainable development.

UN DIVISION ON THE ADVANCEMENT OF WOMEN

The following is from the Division on the Advancement of Women Web site:

> Grounded in the vision of equality of the United Nations Charter, the Division for the Advancement of Women (DAW) advocates the improvement of the status of women of the world, and the achieve— ment of their equality with men—as equal actors, partners, and beneficiaries of sustainable development, human rights, peace and security. Together with Governments, other entities of the United Nations system, and civil society, including non-governmental organizations, DAW actively works to advance the global agenda on

women's rights issues and gender equality and ensure that women's voices are heard in international policy arenas." www.un.org/womenwatch/daw/.

CONVENTION ON THE ELIMINATION OF ALL FORMS OF DISCRIMINATION AGAINST WOMEN AND OPTIONAL PROTOCOL

The Convention on the Elimination of All Forms of Discrimination Against Women and Optional Protocol adopted in 1979 by the UN General Assembly is often described as an international bill of rights for women. The convention defines discrimination against women as any distinction, exclusion, or restriction made on the basis of sex that has the effect or purpose of impairing or nullifying the recognition, enjoyment, or exercise by women, irrespective of their marital status, on a basis of equality of men and women, of human rights and fundamental freedoms in the political, economic, social, cultural, civil, or any other field.

By accepting the convention, states commit themselves to undertake a series of measures to end discrimination against women in all forms, including:

- To incorporate the principle of equality of men and women in their legal system, abolish all discriminatory laws, and adopt appropriate ones prohibiting discrimination against women

- To establish tribunals and other public institutions to ensure the effective protection of women against discrimination

- To ensure elimination of all acts of discrimination against women by persons, organizations, or enterprises.

Countries that have ratified or acceded to the convention are legally bound to put its provisions into practice. They are also committed to submit national reports, at least every four years, on measures they have taken to comply with their treaty obligations.

BEIJING DECLARATION

A result of the Fourth World Conference on Women in 1995, the Beijing Declaration is an international agreement for member states to take action to achieve the equality and empowerment of women. The agreement recognizes poverty and environmental degradation as two critical areas of concern in terms of achieving equality for women. The declaration describes itself as a platform to action for equality.

Major areas addressed by the declaration include:

- Women and Poverty

- Education and Training of Women

- Women and Health

- Violence against Women

- Women and Armed Conflict

- Women and the Economy

- Women in Power and Decision Making

- Institutional Mechanism for the Advancement of Women

- Human Rights of Women

- Women and the Media

- Women and the Environment

- The Girl-child

UN Programs and Offices on the Role of Women

- Commission on the Status of Women

- Committee on the Elimination of Discrimination against Women

- Oversees implementation of Beijing and CEDAW (Convention on the Elimination of All Forms of Discrimination)(through January 8, 2008—the convention is now administered by the Office of the High Commissioner for Human Rights in Geneva)

- Regular Publication of the World Survey on the Role of Women

- Other UN conferences impacting women's rights include Habitat II, Istanbul, 1996; World Summit for Social Development, Copenhagen, 1995; International Conference on Population and Development, Cairo, 1994; UN Conference on Human Rights, Vienna, 1993; UN Conference on Environment and Development (UNCED), Rio de Janeiro, 1992

See also **Volume 3, Chapter 4: The Role of Women.**

References

Braidotti, Rosi et al. 1994. *Women, the Environment and Sustainable Development: Towards a Theoretical Synthesis.* London; Atlantic Highlands, NJ: Zed Books in association with INSTRAW.

Jiggins, Janice. 1994. *Changing the Boundaries: Women-Centered Perspectives on Population and the Environment.* Washington, DC: Island Press.

Shiva, Vandana. 2005. *Earth Democracy: Justice, Sustainability, and Peace.* Boston, MA: South End Press.

UN, Women Watch, www.un.org/womenwatch/.

UN Population Fund, a specialized agency of the UN sponsored the International Conference on Population and Development in Cairo, Egypt from September 5–13, 1994.

International Public Participation: The Aarhus Convention

For further discussion see volume 3, chapter 2.

Agenda 21 and Local Agenda 21 Movements

Agenda 21 is a comprehensive plan of action designed for global, national, and local sustainable action by organizations of the United Nations System, Governments, and Major Groups in every area in which humans impact on the environment. Agenda 21 was adopted by more than 178 governments at the United Nations Conference on Environment and Development (UNCED) held in Rio de Janerio, Brazil, June 3–14, 1992. The full implementation of Agenda 21, the Programme for Further Implementation of Agenda 21, and the Commitments to the Rio principles, were strongly reaffirmed at the World Summit on Sustainable Development (WSSD) held in Johannesburg, South Africa, from August 26 to September 4, 2002.

Agenda 21

Agenda 21 is a 900-page document outlining the goals and responsibilities of human government toward sustainable development. The major areas of Agenda 21 include:

- *Social and economic dimensions*, including addressing poverty, health, and urbanization.

- *Conservation and management of resources for development,* primarily the protection and preservation of the earth and wildlife, and pollution control.

- *Strengthening the role of major groups*, particularly groups that are traditionally marginalized including women, children, developing nations, and indigenous people groups. Nongovernmental organizations, industry, and farmers are also key groups.

- *Means of implementation*, including improving education, incorporating technology, funding for sustainable projects, and national and international tools for decision making.

References
Agenda 21, www.un.org/esa/sustdev/documents/agenda21/index.htm. *See also*, **Volume 1, Appendix C.**
International Council for Local Environmental Initiatives, www.iclei.org/.

Habitat II and the Habitat Agenda (1996)

The United Nations Human Settlements Programme, UN-HABITAT, is the United Nations agency for human settlements. Habitat is headquartered in Nairobi, Kenya, and has regional offices throughout the world. Habitat is managed by an executive director and has three main agencies: the Shelter and Sustainable Human Settlements Development Division, the Monitoring and Research Division, and the Regional and Technical Cooperation Division. It has technical assistance programs directed for the urban poor in 61 nations, with a special emphasis on post-conflict and post-disaster urban areas. It maintains a database called the Statistics Programme that regularly collects data from member countries and cities, as well as the Urban Indicator Programme that regularly collects indicators from more than 200 cities. It tracks Goal 7, Target 11 of the Millennium Development Goals. This goal includes sanitation, safe water, and safe and secure housing for humans. Its mission is to promote socially and environmentally sustainable towns and cities with the goal of providing adequate shelter for all. The UN Habitat Programme actively seeks to reduce urban poverty through the implementation of sustainable urbanization, waste management, and economic development. It provides job training and housing assistance in more than 61 nations and monitors sustainability and the state of the world's cities.

As a result of the second Habitat Convention in Istanbul, Turkey, 1996, the UN adopted its Habitat Agenda, widely regarded as the precursor to the Millennium Development Goals. The Habitat Agenda is the primary political document that resulted from the Vancouver and Istanbul conferences and has been adopted by more than 171 nations, including the United States.

The Istanbul Declaration, or Habitat II, focuses the Habitat agreement and recommended action on sustainable human settlement and adequate shelter for all humans. Signatories also explicitly agree to pursue sustainable production and consumption and to consider the needs of future generations in policy. Habitat II also ties the goals of the Habitat Programme to Agenda 21.

Since 2001, the UN Habitat Programme has issued a regular state of the world's cities report. This report provides a global scorecard on urban issues and an assessment of global progress toward meeting millennium development goals related to urbanization. The recent 2006/2007 report focuses on the difference between urban and rural poor. It reveals that the rural poor often have a better quality of life than the urban poor, but with substantial variation between regions. The report also shows how poverty was reduced in nations that have reduced slum growth in the last decade.

References

Habitat Agenda, www.unhabitat.org.

UN, Human Settlements Programme, Habitat Agenda, ww2.unhabitat.org/hd/hdv10n2/4.asp.

UN, Human Settlements Programme, ww2.unhabitat.org/.

Stockholm Convention on Persistent Organic Pollutants

Some human-generated pollutants do not break down in the environment. Instead, they accumulate in the environment and bioaccumulate in the fatty tissues of animals. Scientific agencies of the UN have concluded that many of these types of pollutants, called persistent organic pesticides (POPs), pose significant risks of harm to human health and the environment. These include disruptions in our endocrine systems, reproductive systems, immune systems, nerve centers, and a range of cancers including breast cancer. POPs can occur in nature, but the vast majority of them are the result of their use by humans in pesticides or pharmaceuticals.

The Stockholm Convention on POPs agreed to limit the use of 9 of the worst 12 such compounds, with significant exceptions for malaria (using DDT to kill mosquitoes). Banned POPs are aldrin, chlordane, dieldrin, endrin, heptachor, hexachlorobenzene, mirex, toxaphene, and polychlorinated biphenyls (PCBs).

The Earth Charter

An ecumenical group of spiritual organizations has also joined to advocate for the protection of the environment and provision of assistance to the poor. They drafted the Earth Charter, a statement of principles of responsibilities based on faith. This group is an independent global organization representing more than 2,500 organizations, 400 cities and towns, global agencies like UNESCO(United Nations Educational, Social, and Cultural Organization) and IUCN (International Union of Conservation of Nature), and individuals. The charter has four basic principles subdivided into 16 smaller tenets. The four principles are respect and care for the community of life, ecological integrity, social and economic justice including the eradication of poverty, and democracy, nonviolence, and peace.

UNITED STATES

President's Council on Sustainable Development

President Bill Clinton appointed an advisory council to make recommendations on how to implement the idea of sustainable development in the United States. The President's Council on Sustainable Development was formed by Executive Order 12852 on June 29, 1993. The council brought together representatives of industry, government, and environmental organizations. The result was a series of meetings and reports over a period of six years, culminating in a final report called, "Towards a Sustainable America: Advancing Prosperity, Opportunity, and a Healthy Environment for the 21st Century, May 1999," which included 140 specific recommendations to promote sustainable development. These recommendations addressed all levels of government, community organizations, businesses, and individual citizens. The recommendations focused on issues of urban sprawl, climate change, urban ecology, and corporate responsibility for environmental and community consequences.

This council was disbanded under the Bush administration and the recommendations were never followed. It remains to be seen whether the Obama administration will renew this commitment.

U.S. Legislation Requiring Public Participation

Many environmental laws in the United States require public participation, such as the Clean Water Act, the Clean Air Act, and the National Environmental Policy Act. The notice given to the public is often ineffectual and late. Much of the public participation occurs after the environmental decision has been made. In terms of sustainability, much of this type of late and ineffectual public participation is not useful to making decisions that affect natural systems on which future life depends. The cost of public participation often is too expensive for poor people to be meaningfully included under these laws.

Reference
Collin, Robert. 2006. *The Environmental Protection Agency: Cleaning Up America's Act.* Westport, CT: Greenwood Press.

The U.S. Environmental Protection Agency: Public Involvement Policy Overview

The U.S. EPA has a legal commitment to require public participation. The president and Congress try to strengthen those legal requirements depending on their own environmental goals. With the rise of public access to more environmental information, many agencies are challenged to meet community demands for more involvement and more power. EPA underscores the need for more early and meaningful public involvement. It wants to ensure that environmental decisions are made with an understanding of the interests and concerns of affected people. It promotes more techniques (discussed later) to create opportunities for public involvement with EPA decisions. The EPA applies these procedures to many of its operations, such as making rules for environmental regulation, permit issuance, renewal and modification, selection of plans for cleanup of hazardous waste sites, and many others.

The EPA may strengthen its policies, but that is no guarantee that states and localities will do so; and for sustainability, that is where it matters most. The alienation of land use planning from environmental planning in the United States also alienates the EPA from on-site decision making, in many instances. The EPA itself is an agency made up of more than 18 mission statements from Clean Air and Clean Water, to Pesticide regulations and environmental impact statements. The EPA has different roles in these decisions, but usually is the convening stakeholder. It can partner with national projects and community groups, develop recommendations to EPA, and sometimes develop binding agreements. As a decision –maker, the EPA exchanges large amounts of important environmental information and develops recommendations. As a capacity builder, the EPA exchanges information, develops recommendations, and can develop binding agreements.

In most public participation processes in the United States, the government is the convening stakeholder. Most public participation consists of meetings and information exchanges, with a very small part of any one potentially impacted population getting notice or actually getting involved. Some communities are so concerned about the public health effects of accumulating, uncontrolled and unmeasured environmental impacts they monitor the environment themselves. Their "notice" is their own observations, scientific and otherwise, of changes in the environment. Because of this groundswell of community concern and greater knowledge of what public participation opportunities exist, the range of public participation tools that meet minimum legal requirements has expanded. The growing range of public participation tools is a good development for sustainability decision making. These tools will need to be embraced as in the model plans of land use at state and tribal levels.

Environmental Assessments (Environmental Impact Statements)

Most of sustainability policy hinges on environmental assessments of the human impacts on the environment. Assessments of the environment have warned of the degradation of natural systems. Environmental assessment policies determine if and when and what kind of environmental mitigation is necessary. Most environmental assessments relied on by governments are based on some degree of science. There are international standards of environmental impact assessment such as those put forward by the United Nations Social, Educational, Social, and Cultural Organization.

On June 27, 1985, the European Council of Ministers adopted a rule that required its members to adopt environmental assessment procedures. Its members include Austria, Belgium, Denmark, Finland, France, Germany, Greece, Ireland, Italy, Luxembourg, The Netherlands, Portugal, Spain, Sweden, and the United Kingdom. The procedures include a threshold determination of whether the environmental impacts are significant and are applicable to both government and private projects. The council has a list of projects that require an environmental impact statement (EIS) and another list of projects for which an EIS is discretionary.

Communities do their own environmental assessments, and citizen monitoring of the environment is on the increase. Industries also do their own environmental assessments. There are many kinds of environmental assessments. Models of environmental assessment that ignore, diminish, or underreport actual environmental impacts are inadequate for sustainability purposes. It is likely that many of the currently used environmental assessment models will be the springboard for sustainability environmental assessment models.

THE U.S. MODEL

The first U.S. federal law requiring environmental impact assessments was the National Environmental Policy Act of 1970 (NEPA). Many states have since developed their own state environmental policy acts, and some tribes have developed their own tribal environmental policy acts. An overall critique of the environmental assessment policy context is that it does not cover enough of those human activities that result in environmental impacts. They often fail to measure cumulative impacts or ecosystem impacts over long periods, for example. Some environmentalists want all environmental impacts covered by new policies such as a municipal environmental impact statements. Pushing the NEPA environmental assessment model to cover all environmental impacts that threaten natural systems on which all life depends would enrich it to sustainability levels.

NEPA requires a detailed environmental impact statement for major federal actions significantly affecting the quality of the human environment. The express legislative purpose of NEPA is:

> To declare a national policy which will encourage productive and enjoyable harmony between man and his environment; to promote efforts which will prevent or eliminate damage to the environment and biosphere and stimulate the health and welfare of man; to enrich the understanding of ecological systems and natural resources important to the Nation; and to establish a Council on Environmental Quality.

The underlying purposes of NEPA parallel principles of sustainability. They are to:

1. Fulfill the responsibilities of each generation as trustee of the environment for succeeding generations

2. Assure for all Americans safe, healthful, productive, and esthetically and culturally pleasing surroundings

3. Attain the widest range of beneficial uses of the environment with degradation, risk to health or safety, or other undesirable and unintended consequence

4. Preserve important historic, cultural, and natural aspects of our national heritage and maintain, wherever possible, an environment which supports diversity, and variety of individual choice

5. Achieve a balance between population and resource use which will permit high standards of living and a wide sharing of life's amenities

6. Enhance the quality of renewable resources and approach maximum attainable recycling of depletable resources

A fundamental weakness of NEPA is that the environmental assessment is not a document that determines policy. The environmental assessment is ultimately advisory only. This would make it unsuitable for purposes of sustainability because it would not be action oriented enough. Nevertheless, much of the purpose and goals of NEPA follow many principles of sustainability. The processes and participants, or stakeholders, to the NEPA process would be the first used for sustainability environmental impact assessments.

The purpose of the NEPA environmental assessment is to reduce environmental impacts on the environment when possible; however, economic considerations specifically and legally drive the decision-making process because they are the overriding value in U.S. society. Nonetheless, the NEPA Environmental Impact Assessment process offers valuable ways to garner important environmental information.

There are four main sets of participants to the NEPA process. The first is the lead agency, which is the agency responsible for EIS preparation and for making the decision on the proposed action. The second set of participants is the EIS team. This group can differ widely from project to project, depending on the range of issues. Generally, it is an interdisciplinary group of specialists making scientific observations and decisions around these observations. They are usually scientists, engineers, and institutional planners. Each group member is supposed to be fair and unbiased. Each is supposed to examine every area as thoroughly as the level of significance of the environmental impact dictates. Group members can be employees of the lead agency or private consultants working under the lead agency, as well as agency staff.

Some advocates of greater inclusion in the environmental decision-making process and as a principle of sustainability criticize the NEPA process at this juncture, because there is no citizen or community or environmentalist involvement in the core EIS team. The lack of inclusion, they claim, can cause them to issue a finding of no significant environmental impacts when in fact there are significant environmental impacts. The level of significance of environmental impacts is very controversial under NEPA. Short-term impacts resulting from most construction techniques are not considered significant. The environmental impacts of the mitigation techniques themselves are not considered. Some industry stakeholders argue that simple compliance with air, water, and land environmental laws should be considered part of the mitigation package. Under a regime of sustainability policy, significance would be tied to potential for irreparable damage to natural systems on which future life depends. There would be little to mitigate if the damage were irreparable, but controversies ensue until the evidence almost reaches levels of species extinction, as in overfishing.

The third group that is part of the U.S. NEPA process is the project proponent. This is usually a private developer or landowner, or sometimes an agency. It can also be called an applicant or a sponsor. If the project proponents need money, environmental permits, or governmental approval, they probably have to begin the environmental impact process. They have to provide accurate and complete information about the design, construction, and operation of the proposal. They are supposed to share all drawings, feasibility studies, environmental information, and building designs with the lead agency. The lead agency can request more information or explanations of the information provided. Project proponents often complain about the intrusion into business practices and the amount of time an EIS involves. Many claim it scares away potential investors. Sometimes the process of doing an EIS uncovers previously unknown legal liabilities, such as an illegal hazardous waste site. This also affects investors' perceptions of the risk involved with the project. If there are significant environmental impacts that have to be mitigated, the cost of mitigation could be expensive. These contingent liabilities of time-consuming, unknown environmental liabilities, and cost of mitigation also affect the project proponent, as well as the NEPA environmental assessment process.

The role of the public as a participant is varied and growing under U.S. NEPA. Generally interested parties of the public are allowed to provide input to the lead agency in the scoping process to narrow the issues and alternatives in a draft EIS, and to review the draft EIS. Many communities have felt excluded by this process. The lead agency generally selects which parts of the public are allowed to participate. Communities want more of a say in developing project alternatives and in designing mitigation schemes. Under NEPA, public review participants include private citizens, Indian tribes, other agencies that have expertise or jurisdiction, and interested parties who have requested notification.

The NEPA Environmental Assessment Process

The process begins when project proponents submit their application to a federal agency. At this stage of the process, the application may not indicate all the environmental impacts but show preliminary designs and concepts. If it is a state environmental policy act (SEPA), private or local agency actions may not necessarily be covered. Some states require only lead agencies to submit EISs.

After the proposal is submitted, the lead federal agency determines whether the project is categorically excluded from EIS requirements or if it is exempt from them. Both the federal agencies and state governments have these categories. The policy justification for these categorical exemptions is that these are activities that do not usually cause significant impacts and therefore would not require an EIS. Many environmentalists and sustainability proponents, however, argue that this is not always the case. For example, Community Development Block Grants were largely categorically excluded from federal EIS requirements under NEPA. These funds went to many programs in urban areas that had direct and indirect environmental impacts that were not counted or assessed. An adequate platform for a sustainability policy would need to count environmental activities without categorical exclusions or exemptions.

The next step in the U.S. NEPA process is to make a threshold determination of whether there are significant impacts to the environment because of the project. This is a highly controversial and litigated area of the law. Threats to endangered species, wetlands, historical and cultural areas, and controversy itself can trigger a level of significance that requires the EIS. Issues of environmental injustice and racial disproportionality can be a significant impact. If the lead agency is uncertain about whether there will be significant impacts on the environment, it performs an environmental assessment, or EA to determine the potential for significant environmental impacts.

There is substantial community concern and sustainability criticism for this step of the process. Communities are not given notice of this EA or even the project application. They have no opportunity to say what they think the environmental impacts of the proposal would be to them and their environment. The EA is often limited to long-term direct impacts on the study site alone. Some claim the study area is too small to measure ecosystem impacts in most projects, and that the environmental impact study area is manipulated to decrease environmental liability and significance of environmental impacts. For example, in a site where a federal courthouse was proposed, a leaking underground storage tank from an abandoned gasoline station was found. If it were included in the study area, it would have shown a plume of petrochemical pollution from the leaking underground storage tank through the soil, to the water table and to a nearby river. The cost of cleanup and mitigation of this hazard would be expensive and time consuming. The study area was redefined to exclude the abandoned leaking underground storage sites. Although the community resisted because it wanted to get the site cleaned up, the federal government, and its EIS process, expressly preempts state and local environmental laws.

If the lead agency finds no significant environmental impacts, it issues a finding of no significant impacts, or FONSI. This is often the first notice the community receives about a project in its midst. Some communities wholeheartedly endorse any economic development despite environmental consequences. Other communities express shock and outrage at their lack of inclusion on the threshold issue of significant impacts. Environmentally burdened communities are especially sensitive to late notice and exclusion from decisions that directly affect them. Many environmentalists express concern that a more thorough study was not performed and that ecosystem and cumulative effects were not included. Sustainability proponents find this stage of the process lacking because of lack of inclusion and because of lack of ecosystem or biome-based study. The U.S. EPA's Council on Environmental Quality defined ecosystem as:

An ecosystem is an interconnected community of living things, including humans, and the physical environment with which they interact. Ecosystem management is an approach to restoring and sustaining health ecosystems and their functions and values. It is based on a collaboratively developed version of desired future ecosystem conditions that integrates ecological, economic, and social factors affecting a management unit defined by ecological, not political, boundaries.

Interagency Ecosystem Management Task Force, in Council on Environmental Quality, Twenty-fourth Annual Report, 1993.

If the lead agency does find significant environmental impacts, then under the U.S. NEPA process they issue a notice of intent (NOI) to prepare an EIS statement. The NOI is published in the Federal Register. This publication comes out daily and is issued to federal depository library institutions. It constitutes public notice of agency actions such as rules, regulations, and EISs.

From here, the public scoping process begins. Scoping is an important part of the U.S. NEPA process because it can determine the actions, alternatives, environmental effects, and sometimes mitigation measures included in the EIS. Different federal agencies involve the public to different degrees in the scoping process. A common complaint, however, is that only those members of the public who agree with the proposal are allowed to participate in a meaningful way. Many environmentalists and most sustainability proponents would consider the actual scope of the EIS to be too narrow and small to be applied to environmentally based sustainability approaches. Agencies have the discretion to choose members of the public. Until communities complained about environmentally unjust disproportionate environmental impacts, few members of the public were engaged in the scoping process. Scoping is done informally and formally by the lead agency, often in close consultation with the project proponent or their consultants. Communities and tribes have felt excluded from the process because they were excluded. The lead agency determines the size and scale of the environmentally affected area. This is a controversial decision because it underestimates environmental impacts according to many environmentalists.

Another important area of public involvement under the U.S. NEPA process is public review of the draft EIS, which contains all the alternatives to significant environmental impacts. The range of alternatives, including the no action alternative, is often very small. The more alternatives considered, the more expensive and time consuming the EIS process can be. The draft EIS can contain important environmental information that could be useful as baseline information in later sustainability assessments.

Public review of the draft EIS is limited under U.S. NEPA processes. Only comments within a period that address certain questions posed by the lead agency are reviewed. Agencies allow little public review from groups that simply do not want the project at all because of environmental impacts. The public is composed of developers, special interest groups, environmental and economic advocacy groups, individual citizens, and other reviewing agencies. The lack of inclusion at this stage of the NEPA process changed in the late 1990s, primarily because of the political pressure of environmental justice groups and the legal advocacy of environmental groups seeking information from ongoing EIS processes.

After public review of the draft EIS, the lead agency reviews all the comments relevant to the significant environmental impacts and alternatives presented in the draft EIS. These are generally published in the *Federal Register.*

EIS: A Done Deal?

Environmental lawyers must generally wait until an administrative agency makes a final decision. This is a legal doctrine called "the exhaustion of administrative remedies." The purpose of the doctrine is to leave specialized and complex areas within the expertise of the administrative agency until the complainant has pursued all administrative avenues, giving the administrative agency an opportunity to self-correct. This can be a long and expensive process, effectively excluding most poor and working people. Interagency appeals processes can take years. The final decision in many cases is not always clear. In the NEPA EIS process, the final lead agency decision is called the Record of Decision and is published in the *Federal Register* long after the EIS is complete. Many environmentalists and communities feel excluded from meaningful participation, leaving their interests unaddressed. Their interests are often in line with sustainability values and approaches, such as preservation of the environment for future generations and the application of the precautionary principle.

Economic Value Prioritized in U.S. NEPA

It is clear that the U.S. NEPA EIS process is laden with a strong economic value directed toward growth. It is unusual for a project to be denied. Many claim that the environmental mitigation measures claimed in the EIS are unenforceable. The whole decision is not mandatory, merely advisory. From a sustainability perspective, the lack of meaningful participation around ecosystem issues poses a challenge. Although the U.S. EIS process is supposed to examine cumulative impacts, and cumulative impacts are supposed to be a significant impact on the environment, in reality, they are ignored because of the time and resources necessary to evaluate and then mitigate them. The increased environmental scrutiny under assessment procedures, however, may dissuade projects with overwhelming harmful effects from even submitting an application. This is not necessarily the case because of the pro-growth assumption of the U.S. EIS process. Nuclear reactors and nuclear waste sites are subjected to stringent EIS procedures in most cases but are eventually given permission to operate, for example.

Another check on the U.S. NEPA process is the amount of time involved. An EIS can take a long and uncertain amount of time. From the project proponent's perspective, this can decrease investment and ultimately profitability. There is always a certain amount of pressure to streamline the process, which often occurs at the expense of public participation. In contested environmental issues, such as timber sales in national forests, industry feels that some public participation is done to help environmental lawyers prepare for their lawsuits once the administrative decision is final.

U.S. ENVIRONMENTAL PROTECTION AGENCY: CHECKS AND BALANCES?

Most federal agency EIS under NEPA must clear the U.S. EPA, which ensures that the EIS is up to minimal standards. This occurs in two stages of the EIS process. In the first stage, the EPA evaluates the adequacy of the draft EIS and places it into one of three categories. The first category is "adequate." The standard for adequate is that the draft EIS sufficiently lists and describes the impacts of the alternatives developed, and that no further analysis is necessary. The next category is "insufficient information." This means that the draft EIS did not contain enough information for the EPA to assess environmental impacts that need to be avoided. It can also mean that new alternatives were identified that would reduce environmental impacts more than those considered. These new alternatives can come from many sources at this juncture in the process. Communities with late notice but strong political power can be one source. Other federal agencies can be another source. It is the decision of the lead agency to decide the scope of alternatives, but the EPA can label it insufficient if it finds an alternative that should be there. This then requires that the final EIS consider the alternative. It does not require that it accepts it. The last category of EPA evaluation of the draft EIS is "inadequate." This means the draft EIS does address significant environmental impacts. It can also mean that new alternatives were introduced that would reduce environmental impacts but were outside the scope of alternatives available in the draft EIS. It requires that the draft EIS be revised and resubmitted for public review as a supplemental draft EIS.

The EPA can also evaluate the environmental impacts of the action in the draft EIS. There are four categories of evaluation. The first category is "lack of objections." The second category is "environmental concerns." This means that the EPA identified environmental impacts that should be avoided. Environmentalists have criticized the EPA for not using this category as a basis for broader environmental concerns, such as sustainability. With environmental concerns, the EPA indicates that they volunteer to work with the lead agency to mitigate the identified environmental concerns. The lead agency does not have to do so.

The third category is "environmental objections." Here the EPA identified significant environmental impacts that must be avoided. In this case, the EPA intends to work with the lead agency to avoid these identified environmental impacts. The last category is "environmentally unsatisfactory." This means that the EPA identified serious environmental impacts that could endanger the public health and environmental quality. These environmental impacts must be avoided and the final EIS must show it. If the final EIS does adequately deal with these environmental impacts, the matter is referred to the Council on Environmental Quality.

REAL PROPERTY ENVIRONMENTAL ASSESSMENTS

The purpose of this type of environmental assessment is to avoid liability for past and present cleanup responsibilities on the site. The cost of clean land is a large and dynamic factor in a real estate transaction in many industrialized urban areas. A complex and controversial environmental policy question is whether an institution that finances the purchase of a site requiring cleanup is also liable for cleanup costs. Potential buyers and lenders interview contiguous neighbors, sample soil and water, and search public land and environmental records. In the United States, some banks do extensive environmental agency research at the federal, state, and local level. If the land is found to be contaminated, a whole other level of more probing, on-site assessments takes place. These are not human health, ecological, or cumulative risk assessments They are assessments of cleanup liability and generally assume low levels of cleanup limited only to that site. They may assess whether they can divide the property to avoid the contaminated portion of the site. Generally, at this level an environmental assessment must include the costs of remediation in the assessment. A traditional market value appraisal of real property does not offer much useful environmental information about the land. An environmental assessment does provide this information by including actual environmental condition of sites and the cost of cleaning them to the lowest possible standard, without assessment of biological or chemical risks.

ENVIRONMENTAL ASSESSMENTS: ARE THEY ENOUGH FOR SUSTAINABILITY?

The state of environmental assessment is dynamic and growing around the world. The Council of European States requires a post-EIS phase to follow through on the state of any environmental impacts and to see if the promised mitigation is working to mitigate the environmental impacts. The EIS process includes a research and development component so that it may learn lessons from the site and see how that site fights into its ecosystem. Technology will soon allow us to monitor every site on Earth, including remote locations such as the poles. It also allows for a broad regulatory potential for the environment, something needed for new sustainable policies. Environmental enforcement models rely heavily on assisting compliance. In the United States, if an environmental wrongdoer simply admits to the charge, the fine is reduced 50 to 70 percent and most fines are not collected. Sustainability will require strong enforcement models that will be greatly aided by complete, real-time monitoring of the environment. The aim of these enforcement mechanisms will be to deter environmental behavior that degrades the environment. Technological improvements in environmental monitoring via satellites, cameras, and monitoring stations (staffed and unstaffed) may set the stage for sustainability policies. These improvements vastly increase the number of observations of natural systems, and that knowledge contributes to a burgeoning understanding of the interrelatedness of global ecosystems. Environmental assessments are asked to do more and more, such as include cumulative impacts analysis and ecosystem risk analyses. Emerging from the growing knowledge base available for environmental assessments, the growing application of environmental assessments to more projects, and corporate reporting of the triple bottom line (social, economic, environmental See discussion in Volume 2) is a new emerging "sustainability" assessment.

The amount of growth of environmental knowledge is rapidly increasing, spurred by the knowledge that natural systems have limitations and that humans have a large effect on these natural systems. This rapidly increasing knowledge base of environmental limitations relies heavily on environmental assessment. That assessment, however, depends on the purpose for which it is used.

SUSTAINABILITY ASSESSMENTS AND AUDITS

The assessment of sustainability depends on the scale of the assessment. An assessment of a building or industrial facility is a small-scale assessment; models can be found for this type of assessment. The assessment of the sustainability of an ecosystem is a large-scale assessment, and few models currently exist for this type. The inclusion of an assessment of equity in either assessment is a relatively new concept.

At the small-scale level of sustainability assessment, which includes the participation of the community, some basic questions generally start the process. The first set of inquiries is about the accuracy and completeness of the information. Environmental information tends to be project or community specific. Questions about building materials; environmental impacts on land, air, and water; and worker and community engagement need accurate, reliable, and complete information to start a solid sustainability assessment. The answers to these questions all occur in a site-specific context of the ecosystem, such as where the project occurs, where the building materials come from, and where the products and by-products go.

These questions may challenge certain values about privacy, costs of providing this information, and profits. Traditional methods of construction, concerns about pollution liability and cleanup costs, and the time these processes require also fuel any early controversies around a sustainability assessment. To the extent the sustainability assessment is voluntary, these objections can be difficult to overcome. Communities are often advised to trust information from other stakeholders to verify it independently. Some of these sources can be in the Toxic Release Inventory, the Security Exchanges Commission filings of a publicly traded company, and the system of environmental and land use permits of the community.

The next set of questions for a small-scale sustainability assessment addresses the interaction of the environmental practices and policies with the community and workers. As we try to decrease the impact on the environment in order to become sustainable, the use of workers may change. Decreases in environmental impact can come from the workers who are closest to the processes. Generally, replacing a toxic chemical or hazardous process with a less toxic substance or process is sought. Sometimes, changing the nature of how the final product is stored or distributed is considered. Ways to recycle, reuse, or adapt current chemicals are also sought. All these changes have implications for the workers and for the community. Difficult questions arise about the potential loss of jobs for workers. Sometimes this is answered in part by having an industrial facility increase its research and development toward more sustainable products and processes. This set of questions can be very specific to a given industrial sector and the type of ecosystem and community it operates in.

A third set of questions about sustainability assessments on a small scale revolves around energy use. Energy conservation practices, carbon footprints, and alternative energy development are questions here. Energy audits that thoroughly review energy sources and uses are required. Generally, the energy efficiency of most equipment increases over time so that the time period that is used to assess the energy efficiency is important. Renewable, onsite energy sources, such as solar or wind energy, may have high initial startup costs but pay for themselves over a longer time horizon. Some utility companies will buy back excess power generated on the site. Some communities will consider the use of long-term renewable energy sources as an indication of the long-term commitment to the community.

In terms of the equity component of sustainability, communities need to be able to inspect the facility in question. Community facilities, industrial facilities, and governmental facilities should be open to community inspections. In the past, many communities relied on government enforcement officials to police against environmental degradation. This did not prove effective. Most government officials do not visit the sites of the facilities they regulate but rely instead on reports from the regulated facilities themselves. Community inspections need to be carefully considered for purposes of a sustainability assessment. Considerations to the information that is needed and the time it takes to inspect a facility are important parameters. Some facilities, such as industrial manufacturing plants, can be complex. Communities will need to have experts with them, and this can be expensive and awkward. Some companies want to protect proprietary information or the privacy rights of their clients. Any experts should assess the range of needed information before the actual site inspection. Having worker representatives on the site inspection group is important because they are knowledgeable about the internal processes of the facility.

Sustainability assessments at regional or ecosystem levels are still in the developing stage. They will pose additional challenges to a participatory citizenry seeking sustainability.

References

Devuyst, Dimitri et al. 2001. *How Green Is the City?: Sustainability Assessment and the Management of Urban Environments.* New York: Columbia University Press.

Gibson, Robert B. et al. 2005. *Sustainability Assessment: Criteria and Process.* London, UK: Earthscan.

National Environmental Justice Advisory Council: Model Plan for Public Participation

Because of concerns about equity and unfairness in environmental decisions, the U.S. EPA established the Office of Environmental Justice in 1992. It also designed a new organizational structure to integrate environmental justice into all of EPA's programs and policies. In 1993, the EPA formed a federal advisory committee on environmental justice. All federal agencies have advisory committees governed by the Federal Advisory Committees Act. This federal advisory committee is called the National Environmental Justice Advisory Council (NEJAC).

NEJAC is the leading edge of equity at the U.S. EPA. This federal advisory committee is inclusionary and multistakeholder focused. The creation of the NEJAC marked the first time a federal agency brought together stakeholders from communities, academia, industry, indigenous communities, environmental communities, and federal, state, and local government groups to solve environmental problems. All individuals appointed to NEJAC are not paid by the EPA; participation is by appointment but is voluntary. (Robert W. Collin, coauthor, was a member of NEJAC.)

NEJAC subcommittees are created for specific topics and meet independently of the full committee. Subcommittees must bring their recommendations through the full NEJAC, which then brings them to the EPA. In the first years of NEJAC, most of the work was directed to how the EPA could more effectively integrate issues of environmental justice into their programs and policies. In the late 1990s, NEJAC shifted its emphasis on specific issues, some of which impact directly on sustainability. These issues include the environmental permitting process, community-based health research models, safe drinking water standards, safe fish consumption standards, and public participation models. It has specifically examined pollution prevention models incorporating policies of source reduction, sustainable development, waste minimization, and alternative energy.

Public Participation Models

The engagement of the public in any environmental decision is an ill-defined and often controversial area. Because NEJAC is focused on fairness and equity, its plan for participation is more suitable for the broader reach of sustainable development. It is therefore more time consuming and intensive compared with current models of public participation.

In preparing for the public participation, it is necessary for the convening stakeholder to co-plan and cosponsor with the community. The convening stakeholder is usually the one with the most power and resources, which is usually the federal agency. It is best if the convening stakeholder already has established cosponsoring relationships with community organizations.

Resources in terms of time and money are also necessary to share the planning roles for solid public engagement. These roles can differ based on the type of decision and the capacity of the community, but at the planning stage they generally include decision making, the development of an agenda, deciding on clear and agreed upon goals, determining leaders for aspects of the plan, and methods of outreach to all other stakeholders. Depending on the issue and the capacity of the stakeholders to understand the issue, a certain amount of time is necessary to educate each group so that they may participate equally. Time is also required to make sure the materials fit the culture and traditions of the area. Time is saved by using a facilitator who is knowledgeable about the local culture and environmental justice issues.

The question of who should participate can be a lengthy one for purposes of equity and sustainability. From the community perspective, community and neighborhood groups, community service organizations, churches, and schools should be included. Other participants include environmental organizations; tribal organizations; local, state, and federal government agencies; colleges and universities; hospital and other medical groups; and industry and business groups. These participants will have important stakeholder representatives that should participate.

The logistics of the meeting or set of meetings is important for public engagement. They should be easily accessible by public transportation. They should provide childcare and access for handicapped persons. They should be in large enough spaces and served with effective communication. Language translators and sign language interpreters should be used. The time of day and the time of the year should allow the community to attend without loss of fulfillment of cultural duties or work. In many poor communities, if work is missed so is pay. There has been much discussion about paying poor people who miss work and do not get paid. Many other stakeholders in current public participation models are required to participate as part of their jobs. Generally, head tables are avoided so that people feel equally included. The community and the government stakeholders share responsibility for the leadership of presentations and discussions.

The NEJAC Model Plan for Public Participation recommends that clear goals be maintained by referring to the agreed upon agenda; however, the agenda is not necessarily a binding document. Topics relevant to the overall purpose of the meeting should be included, especially if the reason they were not included was the lack of notice to an affected stakeholder. A professional facilitator will give a timeline of the overall agenda, including adequate breaks. She will explain how the agenda fits into the overall issue at hand. The facilitator also arranges the contact people for follow-up information, writes up the minutes of the meeting, and distributes all relevant materials before, during, and after the meeting.

The equity component to sustainable development focuses on participation, notice, and an opportunity to be heard. These are challenging aspects to current environmental decision-making frameworks when dealing with poor, marginalized people. The NEJAC model for public participation has led the way for inclusion in these types of decisions. When sustainability decisions reach the scale of ecosystems, these challenges will loom even larger.

Collaborative Model: A Framework to Ensure Local Problem Solving

The U.S. EPA's Office of Environmental Justice has taken the lead in developing collaborative means of solving environmental problems. The types of environmental and public health issues fit well with many equitable aspects of sustainability. They include childhood asthma, farm worker pesticide protection, consumption of toxic fish, low indoor air quality, drinking water contamination, lead poisoning, and other aspects of the disproportionate exposure to environmental degradation. The collaborative approach it endorses uses geographic information systems to correlate environmental stressors with race and income data. Community members, academics, professionals, industry representatives, and government officials try to resolve complex issues without resorting to litigation or other adversarial methods. They try alternative methods of dispute resolution and work hard to achieve consensus across stakeholder groups. They examine the entire range of EPA legal powers—standard setting, permitting, enforcement, information gathering, and delegation to state environmental agencies of all these functions. All stakeholders are brought to the same level of capacity to understand the best way to collaborate with each other.

U.S. EPA INTERAGENCY WORKING GROUP ON ENVIRONMENTAL JUSTICE AND COLLABORATION

In 2003, the U.S. EPA Office of Environmental Justice announced the selection of 15 Interagency Working Group on Environmental

Justice (IWG) Revitalization Projects. The IWG forms a collaboration of numerous federal agencies working together. These projects demonstrate collaborative interagency and multistakeholder partnerships in the area of environmental justice, sustainability, and community revitalization. Partnerships involve two or more federal agencies working in cooperation with state and local governments, tribal governments, community-based organizations, academia, business and industry, and/or nongovernmental organizations. The intent of these projects is to examine lessons for the development of collaborative models that ensure problem solving and sustainable solutions to a range of environmental, public health, social, and economic issues associated with environmental justice.

The Office of Environmental Justice at the U.S. EPA is committed to developing constructive engagement and collaborative problem solving in environmental areas. The development of collaborative models represents a significant contribution to the environmental justice, community development, and sustainable development policy. The IWG Environmental Justice Revitalization Projects are real-life examples of local partnerships seeking to solve seemingly intractable environmental and equity problems. Some key elements of such an environmental justice collaborative model include the following:

- Issue identification, community vision, and strategic planning

- Community and stakeholder capacity building

- Consensus building and dispute resolution

- Multistakeholder partnerships

- Facilitative and supportive role of government at all levels

- Implementation, and evaluation and dissemination of lessons learned

The IWG is composed of representatives from federal agencies and White House offices identified in Executive Order 12898 ("Federal Actions to Address Environmental Justice in Minority Populations and Low-Income Populations," February 11, 1994). The basic authority and policy initiative for the IWG come from this executive order. The IWG's primary goal is to work with each listed federal agency to integrate environmental justice within its programs, policies, and activities. An important policy approach to promote such integration is greater interagency cooperation and coordination within the context of multistakeholder collaborative partnerships that include actual communities. There are major benefits to this approach that became part of the lessons learned for sustainable development. Interagency coordination and cooperation are needed to help identify available federal government-wide resources that are available to meet environmental justice resource needs and opportunities by the stakeholders. In 2000, the IWG funded and coordinated an initial round of 15 IWG demonstration projects. As a result of the success of these projects, a strong, pioneering collaborative model to address environmental justice and sustainable development in a community context developed. The IWG began a second round of such projects, called IWG Environmental Justice Revitalization Projects.

The cutting edge of public participation of poor and marginalized people in environmental decisions is represented by these experimental policy forays into community collaboration. They demonstrate that it can be done. When the focus of environmental decision making becomes larger in scale to an ecosystem level, it is likely these models will be revisited.

Two reports that documented the development of an environmental justice collaborative model are IWG Status Report entitled *Environmental Justice Collaborative Model: A Framework to Ensure Local Problem-Solving* (EPA300-R-02–001, February 2002, www.epa.gov/compliance/environmental and *Towards an Environmental Justice Collaborative Model: An Evaluation of the Use of Partnerships to Address Environmental Justice Issues in Communities* (EPA/100-R-03–001 and EPA/100-R-03–002, January 2003, www.epa.gov/evaluate).

Regulatory Negotiation

Regulations are rules that govern the behavior of social actions. Environmental regulations are rules that govern behaviors that impact the environment. Permits for industry to emit chemicals into the air or water are controlled by regulations. Land use controls by cities and states are regulations. Regulations often come from laws developed by elected legislatures or the executive or refined by the judicial branch of government. A legislature, such as Congress, will pass a bill into law. If the executive branch signs it into law, then an agency of the executive branch must promulgate, or create, rules and regulations to enact the law. The agency cannot go beyond the intent of the legislation or act unconstitutionally or arbitrarily when developing these regulations, or a court will find the regulations void.

Environmental laws and regulations are often controversial and end up in court. Regulations governing clean air, fresh water, endangered species, solid waste, and hazardous wastes are spread across several federal and state agencies. Regulations are fiercely negotiated by lobbyists, trade associations, activists groups, and communities.

Regulatory negotiation is also called negotiated rule making, or "reg–negs." Some refer to it as policy discussions or dialogues. It includes some of the public along with the group of regulated parties to work with the government agency making the regulations. Facilitators are often used to bring together some type of consensus as to the regulation. Government agencies support this process because they argue that if the regulated parties agree to the regulation, it will be easier to implement and enforce. Some of the issues are very complex. Some argue against regulatory negotiation because they are too complicated, numerous, and time consuming for communities to be meaningfully involved. The affected parties often want the least regulation at the least cost with the least oversight and the least enforcement. One example is the development of cleanup standards for land contaminated by industry. The community wants it clean enough to build schools and houses, but industry considers this too expensive and wants to clean it up only for industrial reuse. Another argument against regulatory negotiation is that there is no one there to speak for the environmental or ecosystem impacts. Some argue this thwarts the will of the people through their elected representatives that passed the law.

Traditional regulatory negotiations do not offer much potential for sustainable development because they often exclude community and environmental stakeholders. Although government agencies and regulated parties, especially industries, prefer them, they do not provide much outreach or capacity building to affected environments or communities.

STATE AND LOCAL GOVERNMENTS

The Role of States in the Untied States and Europe

Many nations are divided into states or regions. Even the supposedly monolithic centralized economies of the United Socialist Soviet Republic had many states or regions. The relationship between different levels of government is called "intergovernmental relations." The relationship of states to a federal or central authority or national government is often how resources and revenue flow from the national government back to states. There are many financing mechanisms, such as matching grants, loans, and technical assistance. In the United States, the U.S. EPA helped many municipalities with waste treatment facility development, often via states. State agencies often have the power to regulate environmental issues delegated to them from the national government as long as they meet the national standards. If the states choose not to offer the environmental policy, then the federal government can come in and operate it. Because states like to politically and economically control what environmental permits they issue, they tend not to relinquish control to the federal government. In the United States, the relationship between state environmental agencies and the regional offices of federal agencies that fund them can be found in the memoranda of understanding between the two entities.

State environmental agencies face new challenges and encounter unique opportunities with the development of sustainability. With traditional environmental policy, states were to enforce environmental laws and assist industry with compliance. The concern is that states may not want to vigorously enforce environmental laws if these laws chase away economic development. If failure to enforce environmental laws results in a degradation of natural systems on which life depends, however, then an important aspect of sustainability is violated. Under a sustainability decision-making structure, new values could supersede older values of profit. State agencies are in a unique position to facilitate sustainability because they are repositories of what environmental information does exist and often have the power of the national government to carry out environmental policies. State agencies have been powerful, generally silent stakeholders in many environmental controversies. New community constituencies important for sustainability have called them out.

Environmental Federalism in the United States and Europe

In the United States and Europe, the authority for making environmental policy and law is divided between the national authority and the regional or state levels. In Europe this is the European Union. For purposes of sustainable development and participation, important issues of environmental federalism become important. How are new regulations developed, monitored, and enforced? When can states enact environmental regulations more stringent than national requirements? These questions became important recently in the state of California. Under the Bush administration, the U.S. EPA did not allow California to enact more stringent air pollution requirements. With the election of President Obama, this restriction was lifted in 2009.

U.S. Environmental Federalism Background

The rise of a central authority for making environmental rules developed quickly after the formation of the U.S. EPA in 1970. A rapid succession of federal pollution laws quickly developed national standards, which expanded to many other environmental areas such as endangered species, surface mining and hazardous materials management, pesticides, and other areas. Many environmental activists distrusted states to develop appropriate regulations that would protect the environment, and even if they did so, they did not believe that states had the capacity or political will to enforce the laws. Many states were hostile to environmental issues because they represented "liberal" politics or because they threatened profits or jobs.

Background of European Environmental Federalism

Environmental federalism in Europe began later than in the United States but has gone further in terms of sustainability. The initial document establishing the European Community (EC), the Treaty of Rome, did not cover environmental issues. Between 1973 and 1983, more than 70 environmental policies were developed by the EC. In 1987, the Single European Act authorized community environmental directives.

The 1987 act was intended to prevent the race to the bottom, or allowing markets to compete with each other for profit based on decreased environmental regulation or enforcement. The European Union has since moved beyond this policy motivation to pass directives that increase the quality of life, preserve ecosystems, and pursue sustainable development.

Roles of States in Sustainability and Equity Today

States are still a major part of most environmental policies. They are closer to the people and to the environment. New areas of environmental policy and activism like sustainable development often occur first at the community or state level. Environmental monitoring and enforcement are often most effective closest to the ecosystem and the pollution. The tension between federal and state approaches to both environmental policy and sustainable development is high, with many controversies. This tension is increased when international bodies, like the United Nations, push for sustainability via treaties. Tension is also increased when the international bodies recognize nongovernmental organizations that can circumvent both state and federal governments. One area where this occurs that affects sustainability and equity is climate change.

References

Rechtshaffen, Clifford, and Denise Anotoli. 2007. *Creative Common Law Strategies for Protecting the Environment.* Washington, DC: Island Press.

Revessz, Richard et al., eds. 2008. *Environmental Law, the Economy and Sustainable Development: The United States, The European Union, and The International Community.* Cambridge, MA: Cambridge University Press.

Environmental Federalism and Climate Change: U.S. and European Approaches

The U.S. and European approaches to international calls for reduction of greenhouse gases to slow down global warming and the rate of climate change are different. Greenhouse gases in the United States are basically unregulated and kept that way by the federal government. In the early 1990s, the United States worked to establish the Kyoto Protocols, but never ratified them via the Senate. The Kyoto Protocols are directed specifically at greenhouse gas emissions and have the force of law. The Bush administration never supported it. It also refused to engage in any regulations whatsoever that would cap carbon emissions. Its major initiative was a voluntary request to polluters called the Global Climate Change Initiative. The Clean Air Act would be the main regulatory vehicle to decrease carbon emissions, but Republican-dominated Congresses chose not to adopt any legislation to do so. Both these Congresses and the Bush administration have also resisted any regulation that would have required greater fuel efficiency in motor vehicles, which would have a dramatic effect in reducing carbon emissions, as well as other greenhouse gases. The average mile per gallon on gas for a U.S. car is basically the same as the first Model T car in 1924, about 25 miles per gallon.

States want to fill this void of federal regulation and leadership because they are closer to the problems caused by air pollution. In the United States, they are stymied by a type of environmental federalism that prevents them from doing so. Buoyed and supported by international sustainability efforts and by the new Obama administration, however, states are moving forward with climate change policies. There are now more than 700 state policies that move in the direction of decreased greenhouse gas emissions. Many of these state policies began under the Bush administration and were directly confronting the failure of the Bush administration to exert any leadership on the issue. Some states attorneys general even filed lawsuits against the U.S. EPA on these grounds, especially when the Bush administration revived the licenses of numerous coal-fired power plants.

References

Haroff, Kevin T., and Katherine Kirwan Moore. "Global Climate Change and the National Environmental Policy Act." *University of San Francisco Law Review* 42 (2007):155.

Ohshita, Stephanie B. "The Scientific and International Context for Climate Change Initiatives." *University of San Francisco Law Review* 42 (2007):1.

THE CASE OF CALIFORNIA AND CLIMATE CHANGE

California is a large state with a large, diverse urban population and many motor vehicles. It uses many cars, and its standards for entry into its markets affect manufacturing processes worldwide. In 2006, it enacted the Global Warming Solutions Act. This law requires the state to lower its greenhouse gas emissions to 1990 levels by 2020. In 2002, California passed a law that required a state environmental agency, the Air Resources Board, to develop regulations that achieve the maximum feasible reduction of greenhouse gases emitted by cars and light duty trucks. These regulations apply only to cars sold in California and manufactured in 2009 or later.

The failure of U.S. environmental federalism to embrace regulations for global climate change overwhelms even the largest state attempts to develop its own regulations. The world is concerned about the impact on future generations of increased carbon dioxide emissions from industrialized countries. Without U.S. environmental federalism embracing sustainability and community engagement, a global decrease in carbon emissions is slower. Engagement with local communities around carbon dioxide is important because emissions of carbon dioxide come mainly from combustion processes. These processes emit other gases that directly and indirectly affect the public health of humans and cumulatively degrade ecosystems. Emissions of carbon dioxide alone generally do not cause local ecosystem or human health consequences. The combustion process emits many other pollutants such as sulfur oxides, mercury, particulate matters, nitrogen oxides, and volatile organic compounds. The 2006 California law directly engages the equity aspect of sustainability in its climate change laws.

THE EQUITY COMPONENTS OF CALIFORNIA CLIMATE CHANGE LAWS: A BASIS FOR SUSTAINABLE DEVELOPMENT

The Global Warming Solutions Act passed in California in 2006 includes several strong equity measures. The law supports the development of a trading emissions market that would move emissions away from areas that are already overloaded. These tend to be communities of color or low-income communities. The act also supports broad-based citizen participation, especially in communities with environmental impacts. Participation is an important part of sustainable development because it can empower historically marginalized communities and can give important ecosystem information. This act requires the California Air Resources Board to engage a multistakeholder forum in developing its regulation, including the environmental justice community, business and industry, academic groups, and environmental groups. This includes holding public workshops on any new regulations in the communities most affected. The act further creates an Environmental Justice Advisory Committee with members from the most environmentally degraded communities.

By tying emerging state leadership to both equity and global climate change, California challenged dormant federal environmentalism. With the election of President Obama and the change to an engaged environmental federalism going in the direction of both states and the international community, the stage is set for greater changes toward sustainability.

ENVIRONMENTAL FEDERALISM IN THE EUROPEAN UNION

Unlike the United States, the European Union (EU) is fully engaged with its member states to find ways to decrease greenhouse gas emissions and slow climate change. In the 1990s, the Netherlands, Finland, Sweden, Denmark, and Germany were developing policies to decrease carbon emissions and energy usage that generated carbon dioxide. These were in the form of taxes. The EU tried to pass a tax on such activities for the entire union in 1992, but the United Kingdom opposed it. The purpose of their taxing structure was to bring greenhouse gas emissions back down to their 1990 levels. The EU developed many voluntary polices related to global climate change that applied to its member states. In 1998, it developed a Burden Sharing Agreement that set an emissions target that allowed some nations to band together to meet 8 percent of their 1990 levels reduction goal. In 2000, the EU began the European Climate Change Program with more than 40 voluntary emission-reduction policies that member states could choose. In 2002, the EU adopted a policy requiring each member state to ratify the

Kyoto Protocol, and they have all done so. This requires EU greenhouse gas emissions to be reduced to 8 percent of their 1990 levels by 2012. This also allows member states to band together to collectively average their greenhouse gas emissions. The tax efforts failed to gain traction until March 2003, but they contained long grace periods and exemptions for energy-intensive industries in some member nations.

Many member states of the EU were pursuing advanced policies aimed at decreasing carbon dioxide emissions. Germany developed a policy with a goal of reducing carbon dioxide emissions by 25 percent by 2005 and 80 percent by 2050, all voluntary. Germany also developed green taxes on energy, green building codes, and government supported research into energy alternatives. The British government has also moved forward with a strong set of voluntary measures from every one of 50 industrial sectors to reduce greenhouse gas emissions. Other member nations are following suit and are encouraged to make them mandatory by the EU.

The environmental federalism of the EU is more cooperative and allows for a broader range of policy options that fit each country. Many of these options are still in the voluntary phase, but under Kyoto protocols they may soon become mandatory. Participation rights to global climate change policies in each member state vary greatly. The range of environmental inequity and its reflection of ecosystem degradation in Europe are currently unknown. Many parts of Europe have a longer history of human settlement than the United States. It is likely that the cumulative impacts of ecosystem degradation are more severe in some areas.

Agenda 21 and the Role of Local Government

Agenda 21 is a comprehensive document on sustainable development. Even so, it recognizes that local action is required to create a sustainable world. Agenda 21 describes the goals and responsibilities of local government. These goals include local and state level meetings and the development of local plans of action to meet other Agenda 21 sustainability goals. By 2002, more than 6,400 local governments in 113 countries were pursuing Local Agenda 21-related projects. The International Council for Local Environmental Initiatives plays an important role in providing information for local governments on the development and implementation of sustainability programs pursuant to Agenda 21.

Agenda 21 is administered by the UN Division on Sustainable Development. It underscores the need to develop regulations and their enforcement to the lowest effective level of government. When discussing the most effective level of regulation, most public policy researchers try to match the level of government with the scale of the problem. Local environmental problems are solved locally, state problems at the state level, national problems at the national level, and international problems at the international level. This approach ties causes of public policy problems with their consequences, and facilitates traditional public policy approaches such as cost-benefit analysis. The approach stumbles, however, when dealing with environmental problems and ecosystems that span some or all of these levels. Almost all environmental problems begin locally, and excluding some communities excludes dealing with the root causes of some environmental issues. Applying traditional public policy approaches to sustainable development also disconnects causes of the problem from their consequences and often results in underregulation. That is why Agenda 21 emphasizes the delegation of regulatory authority to the lowest effective level of government and in this way heightens the importance of local government engagement with sustainability.

Local Government and the Precautionary Principle

The precautionary principle says that when harm to the environment or human health will result from development, reasonable measures should be taken to prevent harm, even if the scientific evidence is inconclusive.

The burden of proof is too high on those most affected by environmental degradation. The precautionary principle seeks to make certain there is no risk of irreparable damage to natural systems on all live depends. This essentially shifts the burden of proving that a given project is not harmful to the project proponent. Most project proponents are in a large classification of stakeholders called industry, but project proponents can be community residents, as well as state and local governments.

One of the main objections to the precautionary principle is that it forces industry to assume the risks of the unknown, which many argue should be borne by all. Many point out that in most instances, it is not clear how the precautionary principle is actually applied. This greatly raises concerns about costs. Industry feels that a traditional, science-based human health risk assessment is enough. As to the burden of proving the safety of its environmental impacts, industry strongly objects, underscoring how much scientific testing it already does for compliance with environmental laws and for product safety.

In September 2003, the National Association of County and City Health Officials (NACCHO) passed a resolution recommending a policy using the precautionary principle. The resolution was adopted by some U.S. municipalities, such as Portland, Oregon, and Arcata and San Francisco, California, in their land-use planning processes. Land-use laws must have some basis in protecting the public health, safety, and welfare. The resolution states:

> Whereas, land use decisions may contribute to:
>> Health inequities; an increase in health and safety risks, poor quality housing, unstable neighborhoods, unsustainable ecosystems, and poor quality of life can be created; asthma mortality is approximately three times higher among African-Americans than it is among whites; the elderly and people with disabilities are disproportionately affected by a lack of sidewalks and depressed curbs; and

Chronic disease; more than 25 percent of adults in the United States are obese, and more than 60 percent do not engage in enough physical activity to benefit their health; research has shown that a healthy diet and physical activity can prevent or delay type 2 diabetes; and

Increased traffic congestion, reliance on the automobile, and increased pedestrian and bicyclist vulnerability; commuting stress has increased in recent years, while there has also been a decline in social capital (community connectedness); one pedestrian is killed in a vehicle accident every 108 minutes and injured every 7 minutes; and

Decreased air quality and increased pollution emissions; motor vehicles are the largest source of manmade urban air pollution, and the EPA attributes 64,000 premature deaths per year to air pollution; between 1980 and the mid 1990s, the rate of people with asthma rose by 75 percent; and

Decreased water quality; according to the EPA, soil erosion, and destruction of wetlands threaten surface and ground water quality, which may be drinking and/or recreational water sources; runoff from point and non-point sources pollute waterways, and is exacerbated when the amount of impervious surface in an area is increased; and

Loss of greenspace and land conversion; greenspace provides benefits for air and water quality, as well as for the physical and mental health of people; sprawling development consumes 1.2 million acres of productive farmland per year; according to the American Farmland Trust, land is being developed at two times the population growth rate; and

Inappropriate hazardous materials facilities siting, transportation, and storage; exposure to heavy metals has been linked with certain cancers, kidney damage, and developmental retardation; and areas zoned for hazardous materials storage that contain toxic-waste facilities are often located near housing for poor, elderly, young, and minority residents.

Although not all these justifications will apply to all areas, many of them do currently apply to U.S. cities. After calling for the precautionary principle, the NACCHO resolution suggests three ways to make it work:

1. Integrate public health perspectives and practice (which are based on prevention) into land-use planning.

2. Ensure early, sustained, and effective participation by affected community members in all stages of land-use and zoning decisions.

3. Dedicate more resources to getting public health people involved in land-use decisions through training, development of tools, technical assistance, and other support.

The principle is widely used in Europe and increasingly so in Canada. It also directs governmental agencies to restrict the use of products or compounds suspected of producing health risks. The precautionary principle is articulated in Europe as a measure requiring industry to prove that a product or compound is safe rather than showing it is not harmful. This may just seem like semantics, but the articulation has the effect of creating requirements that are more stringent, which protects the health of the public.

References

Myers, Nancy J., and Carolyn Raffensperger, eds. 2005. *Precautionary Tools for Reshaping Environmental Policy.* Cambridge, MA: MIT Press.

Whiteside, Kerry H. 2006. *Precautionary Politics: Principle and Practice in Confronting Environmental Risk.* Cambridge, MA: MIT Press.

Urban Approaches to Sustainability in the United States

The city and county of San Francisco (Calif.) were the first to adopt the precautionary principle in June 2002. Two municipalities that formally adopted the principle are the City of Berkeley, California, and Lyndhurst, New Jersey.

THE SAN FRANCISCO EXPERIENCE: SUSTAINABILITY IN LOCAL U.S. GOVERNMENT

Both the city of San Francisco and the County of San Francisco adopted the precautionary principle first in the United States. These are wealthy municipalities on the West Coast with highly educated and diverse populations. They are not especially industrial in terms of economic development but do have high concentrations of population. They are nestled between the Pacific Ocean to the west and mountain ranges to the east. The city of San Francisco backs up to a large bay.

The adoption of the first local law, called an ordinance, was preceded by intense political dispute. It is likely to be copied by other municipalities, and the two discussed here did adopt significant parts of their language. Sections of this new law are directly related to developing a policy of sustainability and often do more than adopt the precautionary principle alone.

The local elected body, called the board of supervisors, made specific findings to presage the law. These findings were that every San Franciscan has an equal right to a healthy and safe environment. This requires that the air, water, earth, and food be of a sufficiently high standard that individuals and communities can live healthy, fulfilling, and dignified lives. The duty to enhance, protect, and preserve San Francisco's environment rests on the shoulders of government, residents, citizen groups, and businesses alike. These findings seem obvious to most people in most communities, but the local elected officials in San Francisco made them the basis for the adoption of the precautionary principle.

The San Francisco Board of Supervisors noted that historically, environmentally harmful activities have been stopped only after they have manifested extreme environmental degradation or exposed people to harm. It took special note of DDT, lead, and asbestos. In these cases regulatory action took place only after disaster had struck. The delay between first knowledge of harm and appropriate action to deal with it can be measured in human lives cut short. The board considered themselves a leader in making choices based on the least environmentally harmful alternatives. They specifically rejected traditional assumptions about risk assessment and management. San Francisco already had other local laws that dealt with specific policies and that included the precautionary principle. These laws included the Integrated Pest Management Ordinance, the Resource Efficient Building Ordinance, the Healthy Air Ordinance, the Resource Conservation Ordinance, and the Environmentally Preferable Purchasing Ordinance.

San Francisco elected officials sought to integrate all these environmental laws into one policy approach under the broad rubric of the precautionary principle. They also sought to have a more just application of these environmental laws. Their specific goal was to be sustainable by creating a sustainable San Francisco Bay area environment for present and future generations.

The city included a focused discussion of the potential solutions, problems and limitations of science in their findings that presaged the precautionary principle ordinance. It found that science and technology are creating new solutions to prevent or mitigate environmental problems. It also found that science is also creating new problems with unintended consequences. In its view the precautionary principle is a policy to help promote environmentally healthy alternatives while preventing the negative and unintended consequences of new technologies. To do this, it uses a form of alternatives assessment based on environmental impacts. This assessment is based on the best available science. The alternatives assessment examines a broad range of options in order, with different effects of different options considering short-term versus long-term effects or costs. It then evaluates and compares the adverse or potentially adverse effects of each alternative, assessing options with the fewest potential hazards. This includes the option of doing nothing.

Another key principle of sustainability, transparent environmental transactions, is applied to the alternatives assessment under the precautionary principle ordinance. The alternatives assessment is a public process because, in the view of local elected leaders, the public bears the ecological and health consequences of environmental decisions. San Francisco's local leaders seek to fully engage the community in meaningful ways to the alternatives assessment process because they found that the final decision is more robust by broadly based community participation in alternatives assessment, as a full range of alternatives are considered based on input from diverse individuals and groups. The community should be able to determine the range of alternatives examined and suggest specific reasonable alternatives, as well as its short- and long-term benefits and drawbacks. This assumes the community has the capacity and time to do so. Citizens are assumed to be equal partners in decisions that affect their environment.

The local elected leaders of San Francisco based their findings for the new precautionary principle on the goals of a future where the city's power is generated from renewable sources, when all waste is recycled, when vehicles produce only potable water as emissions, when the San Francisco Bay is free from toxins, and when the oceans are free from pollutants. These goals are specifically founded in a hope for sustainability. They intend to use the precautionary principle as a means to help attain these goals and to evaluate future laws and policies in major areas as transportation, construction, land use, planning, water, energy, health care, recreation, purchasing, and public expenditure.

The San Francisco Board of Supervisors was also cognizant of the human behavior changes that will be necessary to live sustainably. It found that the precautionary principle as a policy approach will help San Francisco speed this process of human behavioral change by moving beyond finding solutions for environmental degradation to preventing environmental harm.

To implement the application of the precautionary principle, there must first be reasonable grounds for concern. This will be a matter of interpretation by city planners and managers. Where there are reasonable grounds for concern, the precautionary policy approach will be to reduce environmental harm by starting a process to select the least potential environmental threat. This process is made of principles to guide the implementation and application of the precautionary principle, and these are some of the essential elements of sustainability. The first is anticipatory actions to prevent environmental or human harm. The stakeholders of government, business, community groups, and the general public share this responsibility and duty to engage in anticipatory actions, but these are as yet ill defined. If a particular stakeholder does know that an action will cause harm then there is a duty to prevent it. This duty is based on knowledge, and another guiding principle of the precautionary principle law is an expansive right-to-know provision. The community has a right to know complete and accurate information on potential human health and environmental impacts associated with the selection of products, services, operations, or plans. Unlike many other right to know laws, the burden to supply this information lies with the proponent of a particular action, not with the community or government. There is also a strong duty to do full cost accounting in the assessment of alternatives to a proposed action. This is very extensive and includes raw materials, manufacturing, transportation, use, cleanup, eventual disposal, and health costs, even if such costs are not reflected in the initial price. Short- and long-term thresholds should be considered when making decisions. Although not specifically stated in the precautionary principle ordinance, these could include the cumulative impacts of a particular alternative. At the end of three years, the local environmental agency, called here the Commission on the Environment, will evaluate the effectiveness of the precautionary principle ordinance and submit a report.

The San Francisco ordinance incorporating the precautionary principle as an overarching policy to local land, health, and environmental decisions is a new and foundational development in sustainable development. There are many questions about its costs, benefits, effectiveness, and interaction with other levels of government in the United States. Some view it as the U.S. catching up with European municipalities and their approaches. Others view it as economically unworkable, over-burdensome to some stakeholders such as real estate development, and beyond the capacity of local government and citizens.

THE BERKELEY PRECAUTIONARY PRINCIPLE AS LAW

The municipality, or city, of Berkeley California specifically adopted the precautionary principle in its land use laws, called ordinances. Berkeley is a city near San Francisco. Like San Francisco it has a highly educated, relatively wealthy, diverse population with some industry. A primary employer is the University of California at Berkeley, a leading U.S. research center. The specific purpose in the city of Berkeley in adopting the precautionary principle as an overarching policy is to protect the health, safety, and general welfare of the city. The law does this by decreasing health risks, improving air quality, protecting the quality of the ground and surface water, decreasing resource consumption, and decreasing the city's impacts on global climate change. It intends to implement its precautionary principle law in phases.

The city defines its precautionary principle policy to mean where threats of serious or irreversible damage to people or nature exist, the lack of full scientific certainty about the cause and effect shall not be viewed as a sufficient reason for the city to postpone measures to prevent the degradation of the environment or to protect human health. Furthermore, any gaps in the scientific data discovered by the examination of alternatives will serve as guides to future research on the issue. These scientific gaps will not prevent the city from taking actions protective of the environment. As new scientific information becomes available, the city can review these decisions and make any necessary changes.

The city of Berkeley intends to apply the precautionary principle under a set of guidelines. It will use anticipatory actions to prevent harm to the environment or to human health. It specifically states that government, business, community groups, and the public all share this responsibility to engage in anticipatory actions to prevent such harms. It is also guided by an extensive right-to-know requirement. The community has a right to know complete and accurate information on the real and potential health and environmental impacts associated with the selection of products, services, operations or plans. This is one of the most extensive right-to-know laws in the United States. The city is also guided in its application of the precautionary principle by a requirement that all alternatives be fully assessed, and that the alternative with the least potential impact on health and the environment be selected. It must consider the impact of doing nothing as one of the alternatives. It is also guided by a full cost comparison. It needs to consider long- and short-term costs of any product, including an evaluation of the significant costs during the lifetime of the product. This includes raw materials, manufacturing and production, transportation, use, cleanup, acquisition, warranties, operation, maintenance, disposal costs, and long- and short-term environmental and health impacts. These costs are compared to any available alternatives.

The city's application of the precautionary principle is also guided by a participatory decision process that is transparent and uses the best available information. The city's current sustainable practices are incorporated into the new precautionary principle law. The city is to purchase products or services that reduce waste and toxics, prevent pollution, contain recycled content, save energy and water, follow green building practices, use sustainable landscape management techniques, conserve forests, and encourage agricultural biobased products. The city uses redwood that is certified as sustainably harvested. It will decide how the precautionary principle applies to future actions, based primarily on whether the city manager considers it feasible at the time. The new precautionary principle law also requires that an annual report on implementing actions be produced and available to the public.

The precautionary principle as urban policy in Berkeley is slightly more specific than the San Francisco law. It is reviewed annually, whereas the San Francisco precautionary principle law is reviewed every three years. Because the Berkeley policy is more specific and reviewed more, it may be a better research basis for analysts of sustainable development in U.S. local government. It is likely that state and federal levels of government will examine these laws closely. Some fear that state and federal government agencies will use their preemptive power to usurp these local initiatives before analyses of their effectiveness can take place. Others feel that the momentum of international sustainability movements will prevent the use of the preemptive power of higher levels of government.

Neither of these precautionary principle ordinances were formed in a particular controversy. They were both formed around a hopeful goal of sustainability for the future. Other older and more industrial cities also want to shape sustainable urban policies. These precautionary principles can be formed in the heat of environmental controversy.

LYNDHURST, NEW JERSEY

Lyndhurst, New Jersey, is a municipality in the United States with heavy industry and environmental concerns about cancer clusters. After San Francisco and San Francisco County it was the second city to adopt the precautionary principle into law. In an effort to protect its citizens from the environmental impacts of heavy industry, the mayor and his staff developed a precautionary principle following international definitions and most of the San Francisco model.

The municipality of Lyndhurst, known as townships in New Jersey, adopted the precautionary principle as the policy of Lyndhurst. The city basically adopted the international statement of the precautionary principle, which is that when an activity raises threats of harm to human health or the environment, precautionary measures should be taken even if some cause-and-effect relationships are not fully established scientifically. Although this is a broad principle, Lyndhurst narrows its application through its implementation. All its officers, employees, boards, commissions, departments, and agencies are required to implement the precautionary principle when conducting town business.

Lyndhurst further anticipates implementing the precautionary principle by developing new laws to create a healthier environment. It specifically wants to be a model of sustainable development by creating and maintaining a viable and healthy environment for both current and present generations. Its version of implementation of the precautionary principle is as both a policy tool and overall philosophy to advance environmentally healthy alternatives and remove negative and unintended consequences of modern and future technologies.

Another way Lyndhurst seeks to implement the precautionary principle is through an aspect of environmental justice. It wants to provide every resident with an equal right to a protected and safe environment and avoid disproportionate impacts. It will seek to find the least environmentally harmful alternative in every decision it makes. To use the precautionary principle for a healthy environment, Lyndhurst takes a strong ecological perspective of seeking a high standard of safety in the land, air, and water. It specifically wants to prevent environmental degradation and human health risks before they occur, not after they harm people or the environment. Whenever Lyndhurst is faced with a decision, it will analyze and assess all the alternatives. It intends to assess a wide range of alternatives and consider short- and long-range impacts.

Under this type of analysis, it may get to cumulative impact analyses. As Lyndhurst is an older industrial township, these impacts could be significant. The policy of the precautionary principle is to compare and contrast the adverse and potentially adverse effects of each option specifically noting alternatives with the fewest hazards. There are many environmental decisions with potentially adverse impacts that can include those not scientifically proven or disproved. Lyndhurst is sensitive to potentially hazardous activities and will ask if the activity is even necessary under the precautionary principle. It will try to measure how much harm it can avoid in its assessment of alternatives under their precautionary principle policy. This approach is unusual and a foundational development in U.S. sustainability. It places the burden of proving the safety of an activity directly on the proponent, and not on those opposing such activities such as the community or local government. If there is a threat of irreversible damage to the community or the environment, its policy explicitly does not accept the lack of full scientific certainty about the causes and effects as a legitimate reason to delay governmental intervention that protects the environment or the community. The township will revisit decisions as new scientific information becomes available. This may have the effect of increasing scientific knowledge and community-based monitoring of environmental decisions, both of which are necessary for sustainable development.

Like the Berkeley and San Francisco laws, the Lyndhurst precautionary policy has a set of guiding principles. First there must be some reasonable basis for concern. Lyndhurst does have clusters of cancer and a history of heavy industry. Whether this alone is a reasonable basis for the application of the precautionary principle is unknown but would depend on the type of project or decision proposed. Nonetheless, if there is a reasonable basis for concern, then the precautionary policy creates a duty of anticipatory action to reduce harm. The law creates this duty in government, community, business, and the general public. To anticipate environmental or community harm it must first be known. The next guiding principle is a powerful right-to-know law that places the burden of creating the information on the project proponent. It specifically states that the community has the right to know complete and accurate information on the actual and potential health impacts of the alternatives under assessment.

And alternatives are required to be assessed, including the no-action alternative. An obligation is created to select the alternative with the least harmful impact. Alternatives are also to be evaluated under the principles of full cost accounting. This is a comprehensive cost accounting measure that evaluates all the costs including raw materials, manufacturing, transportation, cleanup, waste disposal, health impacts, and any others. It examines the entire lifecycle for the product or service, from cradle to grave. The entire process is to be transparent in that the community and public at large are to have meaningful involvement in finding and selecting alternatives under assessment.

Lyndhurst's precautionary principle policy is based expressly on the preservation of the environment. It creates a duty to enhance, protect, and preserve the environment on every citizen, business, nonprofit, and branch of township government. Under this law the township wants to generate energy from renewable resources, recycle waste, and prevent pollution and toxins from entering its ecosystem. It views this law as providing the vehicle to develop future laws and policies that protect the sustainability of the ecosystem in the areas of land use, urban planning, water, energy, health care, recreation, transportation, government purchasing, and all public expenditures (such as education).

Lyndhurst developed its precautionary principle as law and policy in the context of a heated history of environmental and public health controversy. Court cases and scientists could not resolve this longstanding, simmering, adversarial standoff. Many municipalities are in this position in the United States and around the world. Economic development, capitalism, private property, and scientific postulations of cause and effect are now being directly challenged by communities and their democratically elected leaders. The result is a policy that goes beyond environmental preservation and conservation, beyond minimal protection of the public health, and that rejects current risk assessment models as inadequate. The result is a growing urban policy that is explicitly ecosystem-based, dynamically inclusive of community in meaningful ways, and that reorients local decision making toward precaution.

Local Agenda 21 Process

Local Agenda 21 is an international effort to create a voluntary process of local community engagement. The whole purpose of this engagement is to create sustainable development policies. It tries to create partnerships around the themes of awareness raising, capacity building, and community participation. The premise of Local Agenda 21 is that most environmental problems involve local communities, and that local governments are important in developing support and new programs around sustainability. The explicit objectives of Local Agenda 21 are the following:

- By 1996, most local authorities in each country should have undertaken a consultative process with their populations and achieved a consensus on "a local Agenda 21" for the community.

- By 1993, the international community should have initiated a consultative process aimed at increasing cooperation between local authorities.

- By 1994, representatives of associations of cities and other local authorities should have increased levels of cooperation and coordination with the goal of enhancing the exchange of information and experience among local authorities.

- All local authorities in each country should be encouraged to implement and monitor programs that aim to ensure that women and youth are represented in decision-making, planning, and implementation processes.

Adoption of these objectives is purely voluntary. Agenda 21 requires extensive consultation with all kinds of civil societies to meet these objectives. The consultation processes are to create long-term relationships that raise awareness and allow sustainable development to come from the ground up. It specifically does not describe any endorsed implementation method. It recognizes that most environmental solutions are unique to a given place, and that inclusionary equity-based dialogues will help match any sustainability policies to that place. This allows the sustainable development policy to fit with the people in that ecosystem, and decreases the cost of monitoring and enforcement that are high with top-down command and control types of regulations.

PLACE STUDIES OF LOCAL AGENDA 21

As Local Agenda 21 becomes adopted in different parts of the world, advocates of sustainability are studying how it is working in practice. Land use planning processes are especially important.

One place study was in Hornsby Shire Council in Australia. The local government council formed a Local Agenda 21 committee. This committee focused on community consultation, citizen participation, and citizen empowerment. It then created a community vision for sustainability and assigned it tangible goals. The visioning process also developed eight community themes that were partnered with the businesses there. From there it developed a sustainability indicators project and a logo.

PUBLIC ENGAGEMENT MODELS UNDER LOCAL AGENDA 21

There are three different approaches for structuring participation in the development of Local Agenda 21. They are the priority problem approach, the sectoral or municipal services approach, and the stakeholder or thematic approach. In terms of public participation, the stakeholder involvement approach is often used in application to sustainable development.

The main method of public participation under Local Agenda 21 is to focus on high priority urban environmental and public health problems and engage community stakeholders around these problems. The process is basic and easily facilitated by local government. It starts with the local government presenting information about a particular environmental problem. This itself is often an important step for increasing awareness about the local characteristic of most environmental issues that challenge sustainability. From here, stakeholder workshops and other meeting types are held to prioritize environmental issues. How they are prioritized can be based on many factors such as scale of difficulty and willingness of citizens to work. Overall, the prioritization is based on considerations of sustainable development. After priorities are set, working committees or groups are formed around the highest priority issues. These groups develop a thorough understanding of the issues and present a range of solutions that are based in sustainable solutions. Experts can be brought in to examine the feasibility, financing, and environmental contexts for these solutions.

Public engagement approaches differ by culture, tradition, and environmental paradigms. Some places use a public participation approach under Local Agenda 21 that begins with developing environmental information around sectors of municipal services, such as fire, police, sanitation, or education. Although this is the beginning focus, it includes community stakeholders in the subsequent analysis of the environmental issues, then prioritizes them, and then creates a sustainability plan with specific actions.

Another basic category of approaches for public engagement under Local Agenda 21 focuses on early active involvement of all stakeholders and early consensus on central themes or environmental issues. It is easy to understand and create, but can result in overgeneralized recommendations that veer away from sustainable development.

Local Agenda 21 is spreading rapidly around the world. It relies heavily on the participation of residents in an inclusionary and empowered dialogue. This is why sustainability has such a strong equity component; it helps develop knowledge and understanding of the current capacity of the ecosystem in a way that will help preserve it for the next generations.

References

Stoyke, Godo. 2009. *The Carbon Charter: Blueprint for a Carbon Free Future.* Gabriola Island, BC: New Society Publishers.

The Local Agenda 21—Urban Environment Management, www.gdrc.org/uem/la21/la21.html.

Weak Communication between Local and State Government: An Environmental Problem

Local land use practices and policies do not communicate well with state environmental agencies. This is a weak link in the U.S. environmental decision-making process. State environmental agencies issue permits to industry to emit chemicals into the land, air, and water. Societies with a high value on economic development through industrialization will place a priority on the speed of the transaction because time can erode profit. Industry is concerned about the time it takes the government to process permit applications, renewals, and modifications while at the same time resisting self-reporting or outside monitoring of some environmental information. Local government land use decisions regarding industry is often in the same time crunch when industrial economic development is sought. Local land use policies and state environmental policies conveniently do not share much information about prospective new development, allowing industry to play one governmental agency against the other.

Because speed of transaction is important for economic development, time-consuming and troublesome stakeholders that could stop a project are not included. Or, if the law requires some type of notice to the community, they are included in ways that are not meaningful. Industry will face a big challenge from the equity component of sustainability because this type of decision making is not inclusive and does not accurately represent past, present, and future environmental impacts. When developing a public participation plan, it is important to plan for capacity building of some stakeholders even if it is time consuming.

The environmental dimension of sustainability will greatly expand the period of decisions because so little is currently known about measuring, monitoring, and restoring human habitation to an ecological balance. The traditional, narrow approach of simply matching issues with stakeholders offers a framework useful for equitable decision making under sustainability, but with greatly expanded time requirements. It is likely that this will be an expensive cost and will increase the risk a project is denied or developed with such costly mitigation requirements that no profit is possible.

Regionalism: Urban Sustainability in Mega Cities

See Chapter 1 in this volume for a discussion of Regionalism.

Local Government and Land Use: U.S. Land Use Primer

Land is central to community concerns and to any realistic attempt to pursue a policy of sustainability. Internationally, Local Agenda 21 is explicitly directed to developing awareness and policy at the local level of government around sustainability. It should come as no surprise that how land is used has a direct, persistent, and cumulative impact on the environment. Yet, in the United States how land is actually regulated is a patchwork of politics, law, and mystery. Historically, land use control had little to do with anything remotely environmental and everything to do with the preservation and enhancement of private property values. As environmental concerns have mounted in U.S. urban areas, many of these communities have reinvigorated the land use planning process with environmental innovation and a sustainable focus.

Land use and the control of land are important to any consideration of sustainability. Land is part of virtually every human activity, and it is an absolute requirement for any form of production or development. Human population growth, large increases in waste and pollution, and patterns of capitalism have increased the simple basic need for space. The need for space generally translates into a need for land. Increased efficiencies in transportation and telecommunications have further increased the demand for space by decreasing the need to be near other social and financial institutions. Given the increase in demand for land and the decreased need to be close to any one central location, it is the usual real estate practice to seek out low-cost land.

Land may cost less than other land for a variety of reasons. Land may cost less because it is polluted, near undesirable land uses, or near marginalized populations. Recent rapid increases in the amount and types of waste plus human population increases all greatly increase the pressure to find low-cost land. This greatly increases the overall impact of humans on the land and develops into sprawl.

Land use controls and laws were developed to help preserve property values by separating out nuisance or dangerous land uses, such as pig farms and cement factories, from residential uses. In this way, land use controls can help create wealth over time in a community. Land use control identifies those land uses allowed or permitted, and those that are not. Land use law is the process by which land is allocated for various uses through the recognition and exercise of private property rights and government to influence land. In the 1910s, as the law of nuisance failed to keep pace with the industrialization and health issues of U.S. urban life, municipalities passed ordinances controlling land use. Land use focuses on the built environment, and generally aim to preserve and develop the built environment. Some cultures would preserve the unbuilt environment as cultural preservation.

The history of equity in U.S. cities is replete with environmentally degraded landscapes and unfairly treated people. The footprint of European colonization, African slavery, and African American and immigrant segregation created the groundwork of the modern city. Issues such as the provision of basic municipal services like fire protection, police security, and sanitation have all been tainted by express racism in their formation and development. Some argue that additional municipal services, such as parks and recreation, recycling, and public works are also tainted by racism. Municipal service discrimination by race is a matter of intense controversy, and after much federal and state litigation, legislation, and monitoring, it is now a matter of law and is illegal. These practices are hard to prove in court, however, and many poor communities cannot afford a lawsuit. As sustainability emerges as a social policy, the inclusionary component will include these very populations. Sustainability generally will include a much more comprehensive approach to the ecosystem.

A quick look at early U.S. urban development reveals a degraded landscape, even before the advent of the car. Some claim this facilitates the twin dynamic of human and environmental degradation occurring together, thus contributing to its large impact in creating the footprint of the modern city. When this is true, it is an important lesson from the past for the development of sustainable policies. It can also be a difficult and painful process, because many values have changed since U.S. cities were started. The history of land use in what is now the United States can be traced to the colonization efforts of Western European nations. Populations of roughly 16 million indigenous people were systematically extinguished to about 3 million people. Land, life, and livelihood of indigenous people were taken and often destroyed.

As the colonies began to grow, they imported African slaves to provide labor for further use of the land to grow indigo, cotton, and tobacco. In the United States, slaves were used for the express purpose of exploiting nature, as opposed to their being prizes of war. As such, U.S. slavery did not countenance families, native religions, knowledge, language, or any human aspect of a slave's existence apart from work. Even after the Emancipation Proclamation issued by President Lincoln freed the slaves, racial discrimination continued in all municipal services, state and federal welfare programs, education, housing, and employment. These were often called "Jim Crow" laws. As African Americans migrated north in search of freedom and equality, racial discrimination continued in housing, education, and employment. The footprint of slavery and Jim Crow created much of the current U.S. landscape of accumulated and unknown pollution, large waste sites, and environmental racism.

Land use planning was slow to control these dynamics. Before land use planning was legal, around 1920, cities relied on nuisance laws to control the environment and the use of land. A nuisance land use action was brought to court generally to stop an activity that was interfering with the property rights of adjoining landowners. On rare occasions, these laws are still used today, sometimes successfully. A private nuisance is a harmful act that interferes with the landowners of adjoining property. A public nuisance generally affects a wider area or the public at large.

Nuisance is an awkward and ineffective land use mechanism. Court decisions may not necessarily apply to subsequent purchasers of the land. Environmentally, there is nothing to protect the environment if no one sues in court. These laws require private parties as litigants, adversarial postures that spill out of the courtroom into neighborhood relations, and societal resources in terms of time, energy, and access to power. The nuisance legal jurisprudence was inconsistent and developed few applicable doctrines, especially concerning equity issues in sustainability. Uncontrolled nuisances, such as brick kilns and dumps, threatened everyone's quality of life and environment. As the first wave of suburbanization and car use boomed in the 1920s, people became aware of these unhealthy land uses and some of their effects. Los Angeles limited brick kilns to a three-mile square area, and this law was upheld in court in 1915. Boston and New York City enacted building height restrictions in 1916 and 1919, respectively. Land use zoning was upheld as a legitimate exercise of state power, and not a violation of the Takings Clause of the Fifth Amendment of the U.S. Constitution.

ZONING: A PRIMARY LAND USE TOOL

One of the main land use tools is zoning. Zoning classifies land into permitted uses, such as residential, commercial, and industrial. These categories are further divided into allowable categories of land use. For example, residential land uses may include a category of permissible land uses of single-family detached houses only and another for multifamily rental housing. Modern zoning ordinances can be complex. Some have special "overlay" zones that are land use restrictions beyond the base land use code. The borough of Manhattan in New York City has at least 42 overlay zones.

FLIP FLOP ZONING

Land uses of urban land that predated the zoning law are called "prior nonconforming uses" and generally allowed until the ownership of the property changes. If it is a particularly offensive prior nonconforming use, such as a billboard, then a city may choose other methods to expedite termination of the land use in question. Zoning can be manipulated to create prior nonconforming uses. This is called "flip flop zoning" and is very controversial and sometimes illegal. The zoning is changed to one use for a short time, or flipped. Then it is changed back to the original zoning, or flopped. This can allow land to be developed for a particular use that was not otherwise possible. Flip-flop zoning generally occurs in very political environments such as city councils. Regular zoning law development is required to be preceded by study, public notice, public involvement, and community engagement.

Land Use Variances and Special Use Permits

There are two basic ways in zoning to avoid a particular land use restriction. Both involve decisions by local land use bodies.

The first is a variance. A variance is generally obtained by asking the local government to allow the property owner to use the land in a way not in compliance with the applicable zoning. Legally, the landowner must show that the whole zoning scheme is irrational as applied to him. A variance runs with the land; that is, subsequent owners of the land have the same right as the seller to use it. It requires a submission to the Citizens Planning Advisory Committee. Citizens Planning Committees are supposed to thoroughly examine all major development proposals and make a recommendation to the first elected body of government—usually a city council or county board of supervisors. Citizen Advisory Committees are composed of appointed individuals from the community. Their recommendations are advisory only, and not binding. They have been criticized as representing a small segment of the citizens, because they tend to represent property owners, not renters. They have also been criticized as overrepresenting real estate industries. Citizens Planning Committees have also been virulent environmental battlegrounds for many controversies.

The second technique to avoid land use restrictions by zoning is to ask for a special permit. If granted, this permit runs to the current owner only, not with the land. This permit usually does not go through the Citizens Advisory Committee, but to the first elected body of government via a planning department if one exists. Currently, many small, mid-size, and large U.S. cities are rife with many variances and special use permits that erode the basic integrity of the land use planning process. Most of the exceptions are to allow higher density residential development or greater industrial use, both with greater environmental impacts than the original land use plan contemplated. Most of the exceptions fall on marginalized urban populations that would now be included as part of the inclusionary component of sustainability.

In terms of early implementation of zoning in U.S. cities, when it was first applied it was to exclude people, not land uses. It was explicitly used to exclude Chinese, Irish, Italian, and Jewish people. At first, it was not thought possible that African Americans, indigenous people, and Hispanic people would even consider living in the white suburban enclaves of the 1920s. When marginalized people were not excluded by zoning, they were excluded with racially restrictive property covenants. These are promises owners make when they buy real property, or land. They are usually contained in the deed. They run with the land and were enforced by the courts until the 1950s. Many modern deeds still have old racially restrictive covenants in them; they are simply unenforceable in a court of law.

LAND USE AND ENVIRONMENTAL JUSTICE: EXCLUSION AS A SUSTAINABLE VALUE?

One of the obstacles to equity components of sustainability in the United States is the historical use of land use practices to exclude certain groups by race, ethnicity, or income. Hazardous and environmentally degrading land uses were allowed and encouraged near these groups and disallowed and discouraged from land occupied by more powerful groups. Some argue it is a function of supply and demand, and that those who can afford to live in safer areas do so, whereas poor people must live in hazardous areas because they can afford no other place. From a sustainable point of view, this poses two problems. One is that some hazardous areas later become desirable to higher income people in a process generically known as gentrification, and they become exposed to the environmental hazardous that preexisted there. Another problem is that environmentally degrading uses can accumulate and spread to other areas in the ecosystem without any regard to land use maps or comprehensive plans. This last problem is aggravated by population increases, land uses and populations that sprawl out over the ecosystem, inadequate environmental monitoring, and weak enforcement of land use laws.

It is difficult to convey the impact of these disproportionate land use patterns on communities and the environment. Yet they are extremely important for sustainable development. One problem is that the time frame of land use change is difficult for any one researcher to follow. Another is that the categories of land use definitions can shift between municipalities or even in the same municipality over time. A third problem for researchers is that technology can change faster than the land use category.

One researcher has examined some of these issues from an equity viewpoint. He examined 31 census tracts in seven different U.S. cities. His main finding was that minority communities have more locally unwanted land uses, and more industrial and more commercially zoned land uses than high-income white communities. In terms of industrial zoning. 13 of 19 low-income and minority census tracts had some land industrially zoned. In seven low-income minority census tracts, more than 20 percent of the land was zoned industrial. Industrial land use classifications often contain the most hazardous land uses, and with them come high levels of truck and rail traffic that can also increase and spread the risk of exposure to the community. Some of the uses under industrial zoning in this study included ammonia and chlorine chemical manufacturing, slaughterhouse, atomic reactors, garbage incineration, and storage of combustible materials. The risk of a catastrophic event, such as an explosion or spill, is greater in industrially zoned areas.

In terms of commercially zoned land, 10 of the 19 low-income minority census tracts had at least 10 percent of their land zoned commercial. In seven low-income minority census tracts, about 20 percent of the land was zoned commercial. This is in contrast with the high-income, low-minority census tracts that had 2 of 12 census tracts zoned for commercial uses, with no census tracts zoned for more than 20 percent commercial uses. Commercial use of land is a broad category and can include many hazardous uses. Here it included metals electroplating, chicken slaughter and preparation, mattress manufacturing, candy manufacturing, and cigar manufacturing. It can also include retail, wholesale, and manufacturing distribution. The common characteristic of commercially zoned land is that it is set up for business, not residence. The number of vehicles and the size of the vehicles are generally larger.

In the United States, low-income minority communities are closer to the areas of environmental degradation than most other communities. These areas of environmental degradation have suffered the same exclusion as the communities, and both must now be meaningfully included to prepare for sustainable development.

References

Arnold, Craig Anthony (Tony). "Planning Milagros: Environmental Justice and Land Use Regulation." *Denver University Law Review* 76, no. 1 (1998):3–153.

Zovanyi, Gabor. *Growth Management for a Sustainable Future: Ecological Sustainability as the New Growth Management for the 21st Century.* Westport, CT: Greenwood Publishing.

Statewide Citizen Monitoring

State agencies do the brunt of environmental regulation in the United States. In terms of intergovernmental relationships, they are often well connected with regional offices of federal agencies but poorly connected with local land use processes, as discussed previously. Statewide issues of water and air pollution increase as the impacts accumulate in the urban areas and spill out via the ecosystem. As sustainability advocates have begun to branch out into state legislatures, some state agencies are taking the lead, such as in New Jersey.

State agencies are closer to environmental problems than federal agencies and often know the dimensions and controversies of them first. Several states have begun to tie equity and sustainability together.

OREGON

Oregon has a long history of environmentalism and environmental controversies. Oregon passed the first bottle bill, requiring refundable deposits on all beer and soft drink containers in 1971. Also in the early 1970s, Oregon became the third state to adopt statewide land use planning. Its engagement with issues of environmental equity began at the executive order level in 1993, when Governor Barbara Roberts formed the Oregon Environmental Equity Citizen Advisory Committee. The Oregon Citizens Committee on Environmental Equity was followed by the Governor's Environmental Justice Advisory Board created by Governor John Kitzhaber on August 1, 1997.

After years of legislative lobbying, OR Senate Bill 420 became law. Introduced by the visionary Senator Avel Gordley, this law was signed by the governor in August 2007. This law substantially changes the relationship between communities and state agencies, and may serve as model for emerging sustainable development models. The law creates the Environmental Justice Task Force that reports directly to the governor about state environmental justice concerns and the progress of state agencies in meeting environmental justice goals. It requires a range of state agencies to address environmental justice issues as part of senior executive management reporting and to include it in an annual report to the governor. The law has an expanded version of natural resource agencies. Even some agencies that were not initially included have volunteered to be part of the process. Many of the state agencies are anxious to show leadership in the progress they have made toward these goals.

The Environmental Justice Task Force is composed of 12 members appointed by the governor. The law requires one appointment each from the Commission on Asian Affairs, the Commission on Black Affairs, the Commission on Hispanic Affairs, and the Commission on Indian Services. Members of minority, low-income, academic, and industry groups are also appointed. The authors of this encyclopedia are appointed as chairs of the Environmental Justice Task Force.

The legal charge of the Environmental Justice Task Force is five-fold. It must first advise the governor on environmental justice issues. It must advise the expanded range of state agencies on environmental justice issues, especially community concerns and public participation processes. It is to identify, in cooperation with natural resource agencies, minority and low-income communities that may be affected by the environmental decisions of the state agencies. It is required to meet with environmental justice communities and to define the environmental justice issues in the State of Oregon.

The law also imposes some new requirements on state agencies. The Oregon natural sources agencies are required to consider the environmental justice implications whenever they are deciding if and how to act on any matter. They are required to hold hearings in the community where these decisions will have an environmental justice effect. This is supposed to be preceded by outreach to the affected public in that community. Each Oregon natural resource agency is required to appoint a citizen advocate position that is to encourage public participation and inform the host agency of the effects of its decisions on communities traditionally underrepresented in the public process. The directors of all natural resource agencies are required to file an annual report with the governor on environmental justice issues.

The Oregon Environmental Justice Task Force shows a strong commitment to address equity in environmental decisions. As such, it provides a good platform for use as a springboard to public engagement for sustainable development.

Reference

Collin, Robert W. "Environmental Justice in Oregon: It's the Law." *Environmental Law* 38 (2008):413.

NEW JERSEY

New Jersey is an older industrialized state in the northeastern United States. A small state on the outskirts of large industrial world cities like New York, it is dotted with waste sites and large industrial uses. The New Jersey State Environmental Justice Order allows citizens to file grievances and then engages all stakeholders in search of a solution. New Jersey's Environmental Justice Executive Order No. 96 (Feb. 18, 2004) is available at www.state.nj.us/infobank/circular/eom96.htm.

The executive order is not a court or alternative dispute type of forum. It makes a series of legislative findings of fact on which to base the legislation. These findings of fact closely describe the U.S. urban context of sustainable decision making. These findings of fact are:

- Communities of color and low-income communities in New Jersey are sited in areas with historically higher densities of known contaminated sites.

- Childhood asthma is increasing.

- Childhood asthma is more prevalent in black and Latino/Hispanic communities.

- The federal government emphasizes environmental justice in its federal policies.

The executive order specifically underscores certain values important to New Jersey:

- A commitment to a healthy environment

- Active public involvement with environmental decision making

- Community capacity building through empowerment from public involvement

- Smart growth

- The importance of cumulative impacts

One important development for creating equity aspects of sustainable decision making is that the New Jersey Environmental Justice (NJ EJ) Task Force can hear directly from affected communities. The task force can also directly act to help solve seemingly intractable environmental problems. The executive order states:

> Any community may file a petition with the Task Force that asserts that residents and workers in the community are subject to disproportionate adverse exposure to environmental health risks, or disproportionate adverse effects resulting from the implementation of laws affecting public health or the environment.

The NJ EJ Task Force is also empowered to follow through on all complaints and petitions. Active citizen involvement in environmental problem description, solution development, and solution implementation are required, with a petition of 50 signatures with at least 25 residents also required. The other 25 signatures can come from workers, teachers, or other concerned individuals. The task force has public hearings, collaborative meetings at local and intergovernmental levels, complete transparency in all transactions, and monitoring of implemented solutions. Quantitative and qualitative information about the site, history, and impacts is actively sought. Petitioners must also assess their own capacity to proceed by answering a detailed questionnaire prepared by the NJ EJ Task Force. This is different from most U.S. environmental dispute resolution procedures, because here the capacity of the petitioner is a factor in the solution. This is an important part of equitable aspects of sustainable decision making. The NJ Executive Order contemplates empowerment of communities by engaging in complaint and solution. It actively seeks out other stakeholders to a particular environmental issue, and will meet privately with any interested stakeholder because some stakeholders fear retaliation if their identity is known. This is true in the case of farm workers in the United States, as well as many other oppressed groups around the world.

The first wave of cases accepted by the NJ EJ Task Force was large urban controversies. Cases that are not accepted are not completely rejected either. The task force keeps in consultation with petitioners and may allow them to submit their petition again in the next application cycle. Because of the unique opportunities presented to state agencies, they are able to focus resources on seemingly intractable environmental problems without litigation or adversarial processes and with collaboration, inclusion, and community empowerment. These are the first steps to solving disproportionate impacts of environmental degradation and perhaps toward sustainability. By giving residents a powerful voice in the environmental problem description and solution, the empowerment can assist in the changes in personal consumption patterns that may occur because of sustainability.

MARYLAND: U.S. STATE POLICY COMBINING ENVIRONMENTAL JUSTICE AND SUSTAINABILITY

Other states, such as Maryland, are combining sustainability with environmental justice. The Maryland Commission on Environmental Justice and Sustainable Communities was first formed by executive order on January 1, 2001. It was signed into law as a statute on May 22, 2003.

The main missions of the Maryland Commission are to:

- Examine environmental justice and sustainable communities for

- Areas that can work together for a healthy, safe, economically vibrant, environmentally sound communities through

- Democratic processes and community involvement.

Unlike The New Jersey Task Force, this process is not resident or community driven. Communities are identified based on the availability of state resources, not environmental or public health need. Once the communities are identified, state agencies can assist communities with technical assistance in the form of cumulative risk assessments, some kinds of public health data, and environmental data. Any fact-finding done by the state agencies is based on both environmental justice and sustainability.

State environmental agencies have a large cast of recreational and sporting stakeholders such as boaters, sport fishers, hunters, trappers, and anglers. They have unique access to data submitted by industrial and municipal permit applicants. There is a strong basis for a collaborative approach because of common data collection goals, common data quality standards, and an expanded range of environmental information necessary for sustainable decision making.

Reference
Roseland, Mark. 2005. *Toward Sustainable Communities: Resources for Citizens and Their Governments.* Gabriola Island, BC: New Society Publishers.

Community-Based Conservation Movements and Sustainability

Human communities are often concerned with their environment. In many places in the world, the first response to urban overcrowding is public health. When the urban population outnumbered the rural population for the first time in the United States in 1920, local communities began to give their local governments the power to zone or otherwise regulate land. This was preceded by the City Beautiful Movement and the Garden City Movement. Both these movements emphasized the role of the environment in aesthetics and public health. They also emphasized the need to make private property interests subservient to the common good of public health. By the early 1920s, many urban areas in the United States were contaminated with human and animal waste mixed in with industrial land uses. The field of public health had grown to the level of making people aware of the spread of disease from crowded urban centers to entire cities and surrounding communities. The rapid development of steel frame construction led to the construction of tall buildings known as skyscrapers. Skyscrapers intensified the impact of the environment by intensifying population and blocking out light and air.

Modern community conservation movements are also concerned with the quality of their environment. They are often focused on avoiding a dangerous locally unwanted land use or community interaction with an industrial facility. Several emerging dynamics have increased the saliency of community conservation movements. One is the increased availability of environmental information via the Toxics Release Inventory. Another is the accumulating amount of chemicals that threaten public health and the environment. A third is the increased impact of technology on the environment. Just as the knowledge about waste, its accumulation, and the building technology around the skyscraper provided the impetus to move local governments to regulate land use in the 1920s, modern community conservation dynamics incorporate the same three influences toward sustainable development. Local communities need more information to pursue sustainable economic development. Communities can often build collaborative partnerships with workers to help make sure old and new industries are pursuing sustainability. In a citizens group, communities educate themselves, advance a common sustainable goal, and negotiate as an organization with industry, as well as local government. Communities often collaborate with other stakeholders, such as churches, schools, and environmental organizations, to conserve and protect the environment.

The engagement of environmentalism with communities also pushes the sustainability agenda. It encourages a conception of community development within that of the carrying capacity of the ecosystem that is home to that community. This can be controversial because it can imply that human communities should be subordinate to ecosystem communities. Some argue that nature is there for humans; others argue that natural limitations on carrying capacity must be honored or sustainability will suffer. Nonetheless, there is an increase in monitoring of systems of nature locally and globally. The climate changes that affect communities increase the policy of monitoring so that the carrying capacity issues cannot be hidden.

References

Blewitt, John. 2008. *Community Development, Empowerment and Sustainable Development.* Devon, UK: Green Books.

Callicott, J. Baird. 1989. *In Defense of the Land Ethic: Essays in Environmental Philosophy.* New York: SUNY Press.

De-Shalit, Avener. 1995. *Why Posterity Matters: Environmental Policies and Future Generation.* London, UK: Routledge.

Tilbury, Daniella, and David Wortman. 2004. *Engaging People in Sustainability.* Gland, Switzerland: International Union for Conservation of Nature and Natural Resources.

CITIZEN MONITORING OF ENVIRONMENTAL DECISIONS

Community-based environmental protection policies develop because people in a given community are often in the best position to address their own environmental problems. A community can be a neighborhood, a village, or a region. Sometimes these regions are biomes themselves. The main idea is that people who live, work, play, and worship all have a common interest in preserving or sustaining their environment. They can take into account local social, economic, cultural, and ecological conditions. These policies also create a sense of local empowerment and encourage long-term community engagement and environmental accountability. This fits into the inclusionary aspects of sustainability.

One of the major goals of residential monitoring of the environment is ecosystem protection. Communities want to make certain that natural systems are healthy enough to provide a range of benefits now and in the future. Application of the precautionary principle of sustainability relies on accurate and complete environmental data, often developed by real-time monitoring of the environment. The services provided by healthy ecosystems refer to benefits to both people and place.

COMMUNITIES AND ECOSYSTEMS

Almost all communities would prefer to preserve the environment if they had a free choice. Parks are one expression of this attempt to preserve the environment. Many residents in cities in the United States and around the world, however, have very little choice regarding environmental preferences. With the rise of more inclusionary dialogues on environmental issues as interest in sustainability increases, these environmental preferences will have more forums for expression.

A common environmental preference that undergirds new values of sustainability is the moderation of natural disasters and human activities. Ecosystems often buffer large natural disasters, such as flooding being mitigated by healthy wetlands. Biomass such as trees and plants can prevent soil erosion caused by floods. The economic benefits of an ecosystem will also attract followers. The interaction between the ecosystem and the economy is extremely important for most approaches to sustainability. The period of cost comparisons affects stakeholder's view of sustainability. Short-term profits of some may conflict with other values attached to the land. Nonetheless, many local economies depend on tourism and focus on outdoor recreation. Urban areas often capitalize on waterfronts and boardwalks.

If tourism is an economic driver or base for a community, the motivation for sustainable decision making is strong. Tourists like to visit nice places with lots of environmental amenities. There is a strong rise in ecotourism. Even without a strong tourist economic base, many economies still benefit from enhancing outdoor recreation. Local demand also creates markets. Many communities around the world rely on the commercial benefit of natural resources. Mining, logging, hunting, fishing, and grazing

FIGURE 3.5 • Urban rivers, such as the San Antonio, are an integral part of nature, too. AP Photo/Eric Gay.

all derive economic benefit from their ecosystems, and so do the towns and communities that spring up around them. A secondary economic impact of natural resource extraction industries is the transportation of the goods, people, and services. Ports, railroad depots, truck-worthy roads, and airports are also tied to the same ecosystem in their economic bases. In addition to the usual list of natural resources, other uses of the environment may develop. Medicines from tropical forests are one emerging example. In all instances, technology is rapidly increasing the efficiency with which remaining natural resources are depleted. One concern environmentalists have with communities used to surviving on a depleted natural resource, like gold or coal, is that they will compromise the environment to survive economically. Ecosystem quality also affects property values. The buffering effect of ecosystems can reduce costs of infrastructure, increase aesthetic appeal, and allow for expression of environmental preferences. *Also see* **Volume 3, Chapter 4: Controversies.**

Most cities are still seeking economic development no matter what the cost to the environment. With economic development come jobs, better municipal services, and an overall better quality of life. Economic development seeks to manufacture something and to sell as much of it to the world as it can to make as much profit as possible. One exception is the service industries of destination resorts. When these destination resorts go green, they often pursue sustainability by decreasing their negative impacts on the environment. Developers of these resorts often have access to more financial capital and environmental expertise than the community or local government.

LORETO BAY, BAJA SUR, MEXICO LORETO, MEXICO: SUSTAINABLE GOLF?

Proponents of sustainability closely examine all claims about sustainability. One of the biggest challenges to sustainability is how to build a large construction project in a sustainable manner and still make a profit. A real estate development called Loreto Bay claims to be the largest sustainable development under construction in North America. It is located in Baja, Mexico, in the small and historic town of Loreto. Its specific goal is to become an international model for how a large construction project can be done sustainably.

Loreto is a fishing and tourist destination overlooking the Sea of Cortez, one of the most biologically diverse marine environments on Earth. Baja is a mountainous peninsula with very little rain, especially on the eastern side near the Sea of Cortez. The available supply of fresh water is a controversial issue. With new large hotels, cruise ships, and population expansion, the demand for fresh water dramatically increases. Some of the hotels have large cisterns under them for their guests. As the water supply decreases, so too does its quality. As waste and sewage have increased into a diminishing water supply, the potential for contamination and public health degradation has increased.

To Loreto Bay, development sustainability means meeting the needs of present generations without sacrificing the ability of future generations to meet their own needs. Developers discuss an intergenerational golden rule where one lives so that children enjoy life as fully as their parents did. Loreto Bay development bases its approach to sustainability on three pillars. The first pillar is environmental sustainability, which means preserving and enhancing the ecosystem locally and globally. Developers claim to do this by measuring and improving business impacts on the environment. The next pillar is social sustainability. This means they will improve the health and welfare of the community and region by providing chances for local jobs and business creation, education, and health. The last pillar is economic sustainability, which means profitability with measurable cost benefits resulting from strategic sustainability policies. To them it means creating wealth for investors, homeowners, local jobs, and new business opportunities. These three pillars form the basis of the sustainability commitments Loreto Bay Development has made.

In the beginning, Loreto Bay Development made specific sustainability commitments. To protect and enrich the environment, developers would produce more energy from renewable sources than are consumed, produce more potable water than consumed, and help create more biodiversity, more biomass, and wildlife habitat. To improve social welfare, they committed to implement a regional affordable housing strategy so that people who work in Loreto Bay can afford to live there. They supported the development of a full-service medical center in Loreto and dedicated 1 percent of the gross proceeds of all sales to the Loreto Bay Foundation. This foundation is designed to help with local social and community issues. In many ways, these commitments represent some of the best practices in sustainable policy implementation. They are also pioneering practices in a profit-driven, highly leveraged, real estate industry.

Because of the concern about water, Loreto Bay claims it will be a water-independent development. This means it will not rely on the local watershed, called the San Juan Basin well field. This watershed supplies the town of Loreto. Loreto Bay Development will get its own water by restoring the two existing watersheds on its land. It will do this by slowing down rain runoff in strategic channels. There is some question about whether this will work.

The buildings of Loreto Bay Development are designed to be energy efficient. The building materials are a combination of recycled Styrofoam, concrete, and compressed earthen blocks. This is very energy efficient, requiring less heating and cooling. Developers also use design features like inside courtyards, fountains, and dome-vented kitchen cupolas to reduce heat. The wood that is used in the construction is certified Bolivian cedar and alder, and urea-formaldehyde free Columbia birch plywood. The wood is certified by the Forestry Stewardship Council, which promises that the wood was harvested in a sustainable manner. This means that the ecosystem is not damaged, the local community and culture are respected, and the local economy is enhanced by the wood.

Construction projects and large developments generally create waste; part of any sustainable construction project is the plan to handle it. Loreto Bay Development plans to us biodegradable soaps and encourage residents to do so. It uses latex paint with very low volatile organic compounds. Waste is divided into three streams at the residence: recyclable, organic compostable, and nonrecyclable. The recyclable and organic compostable waste is to be taken to a waste management center. Some critics of the development do not see enough accountability for waste in the waste management plan.

In terms of environmental enrichment and ecosystem sensitivity, Loreto Bay Development started with a strong vision based on sustainability. It is creating 25 hectares of restored estuaries and planting thousands of mangroves. These two environmental interventions go far to enhance the ecosystem because they provide food at the base of both marine- and land-based food chains. As estuaries and mangrove develop, they provide an ideal habitat for many species such as fish, birds, and snakes. According to calculations, one hectare of mangrove forest produces one ton of commercial fish per year. As part of the environmental enrichment approach, developers are planning to restore two watersheds over the next 20 years. They contend that more water will be absorbed into the ground. This will then result in more flora and fauna. In developing this watershed, they developed a 5,000-acre natural preserve. Local grazing animals often overgraze areas, which can decrease the ability of the soil to hold water. As mentioned earlier, water is a controversial issue. Hydrologists from U.S. universities, real estate investment firms, agribusiness, and governments have a variety of opinions about water quantity and quality. It could be that the two restored watersheds simply drain into a rapidly depleted aquifer.

One unique environmental enrichment activity occurs around the golf course. Most sustainability proponents consider golf courses too wasteful of water and a source of pollution because of pesticide and fertilizer runoff. Many golf carts and lawnmowers still use inefficient gas engines that cause air pollution. At Loreto Bay Development, however, the golf course is composed of a saltwater-tolerant grass called paspallum. Part of the golf course design incorporates bioswales that hold rainwater. Only nontoxic fertilizers and pesticides are used. Most irrigation is done with reclaimed, treated wastewater, a kind of gray water.

Most large real estate development projects are phased. First, phase 1 is built, bought, and sold. Construction loans from banks and other lending institutions are filled with terms punishing late payment or slow work. These large lending institutions are very risk averse and shy away from any potential environmental liability or unusual practice. In a project such as Loreto Bay, the embracement of sustainability did not meet high market expectations. The first primary investor sought and got protection under bankruptcy laws. The bank then owned it and was not knowledgeable about the sustainability principles, and now the Town of Loreto may own it. It is only about one-third built.

One new component to the water issue is the local impact on climate change. If the sea level rises on both the Pacific Ocean and Sea of Cortez side, it could make what is now the peninsula of Baja into an archipelago. If fresh water usage continues to increase faster than it is replaced, the rising sea levels could taint the remaining freshwater sources. The sustainability principles of the Loreto Bay Development may end up serving the town of Loreto, and perhaps all of Baja. If climate change is accompanied with even less rain, the situation could escalate more quickly.

CIVIC ENVIRONMENTALISM

Civic environmentalism is a relatively new concept important to sustainable development and the equity aspects of inclusion. It is a response to command and control types of federal and state regulations. These types of regulations make rules, or commands, that aim to control behaviors that affect the environment. Most environmental laws fall into this category. They do have some room in them for citizen participation, but it often takes place after the main decisions are made. Communities do not feel meaningfully included. Most of these environmental laws do allow for streamlined access to federal courts, but these have been too adversarial and limited to effectuate adequate citizen engagement. Powerful communities have exercised their political and economic clout to prevent or deter environmentally degrading activities, but poor and politically disenfranchised communities have not been able to do so.

Litigation often builds distrust and ill will. Communities seek to solve environmental problems through collaboration, trust, and good will. Adequate, timely, and accurate environmental information has facilitated collaboration and an increase in the capacity to engage in complex environmental decisions. The failure of top-down environmental laws to change behavior enough to matter for sustainable development and the global push for sustainability has also contributed to the rise of civic environmentalism. Civic environmentalism focuses on real problems in real places; even the sources of the environmental problem come from elsewhere. For example, environmental laws rely heavily on risk assessments measures to prioritize policies. Civic environmentalism does not rely on risk assessment, but rather on methods such as Local Agenda 21.

Another complex issue for environmental law is the question of who represents the public interest. Do national environmental organizations really represent the interests of urban poor communities of color? Many of these communities said no, and the early beginnings of civic environmentalism began in the United States as part of the environmental justice movement. These communities did not have equal access to courts and did not feel adequately represented by national environmental organizations. They successfully advocated to decrease environmentally degrading activities; however, many of these activities have to go somewhere. When environmental justice communities successfully stopped their disproportionate exposure to environmental risks, they moved that concern to other neighborhoods who then became more civically involved.

The strength of civic environmentalism lies in its long-term relationships, collaboration, and trust. As local communities become part of a network with each other, and as the international community pushed for sustainability, civic environmentalism will continue to significantly increase.

Reference

Guber, Deborah Lynn. 2003. *The Grassroots of a Green Revolution: Polling America on the Environment.* Cambridge, MA: MIT Press.

LAND TRUSTS

Like private foundations, land trusts are nonprofit organizations that can be a viable funding source for brownfields redevelopment projects. Land trusts exist to protect open spaces and greenways through conservation, land donations, land purchases, and estate planning. Funding for the Seattle Art Museum's (SAM) outdoor sculpture park came in part from a land trust. SAM purchased 8.5 acres in downtown Seattle that included a former petroleum tank farm. Funding for the acquisition was obtained through a private fundraising campaign conducted jointly by the Trust for Public Land (TPL) and SAM. The TPL is a national, nonprofit, land conservation organization that saves land for people to enjoy as parks, community gardens, historical sites, rural lands, and other natural places. A similar group is the Land Alliance, a national organization that is brownfield-friendly and lists each conservation- and environmental-related trust in each state in the United States at www.lta.org/regionallta/northwest.htm.

International Local Government Initiatives

The United Nations, the World Bank, the World Health Organization, and other international organizations focus on nongovernmental organizations to try to encourage local groups and civil societies to participate in environmental decisions. Many communities do not have these groups but do have some form of local government.

INTERNATIONAL COUNCIL ON LOCAL ENVIRONMENTAL INITIATIVES

One international group that focuses on sustainability internationally from a local government perspective is the International Council on Local Environmental Initiatives (ICLEI). The ICLEI is a membership association of towns, cities, municipalities, counties, and the groups that represent them. Its goal is to help build local capacity for decision making around sustainable development by sharing knowledge from its extensive network of locally designed initiatives. It shares this knowledge by doing local trainings and providing consulting on the issue of local sustainable development.

The ICLEI was formed as the International Council for Local Environmental Initiatives in 1990. More than 200 local governments from 43 nations met in New York City at the first World Congress of local governments focused on a sustainable future. This meeting raised awareness of the power of local governments to individually and cumulatively effect social changes necessary for the implementation of sustainable development. This congress recognized ICLEI as the international agency for local governments around the world. So far it has worked with

Aspen, Colorado

Aspen, Colorado is a wealthy mountain community with a strong environmental ethic and growing economy. Much of its wealth comes from winter recreational activities and tourism. It was the first U.S. municipality to evaluate its carbon footprint. It publishes an annual sustainability report that covers transportation, water impacts, air quality, and global impacts. Aspen strives to have an inclusionary environmental planning process and affordable housing. It has its own Global Warming Canary Initiative. According to its 2007 Annual Sustainability Report, the city of Aspen noted increasing temperatures and lowered precipitation. Projections for global warming would reduce and eliminate Aspen skiing by 2100.

The city of Aspen joined the Chicago Climate Exchange, North America's first carbon trading system. Under this program, the greatest greenhouse gas reductions per dollar spent are achieved, because members who can afford to reduce greenhouse gases more cheaply are encouraged to do so and can sell their greenhouse gas reductions to others who do not meet their goals. One of the main ways the city of Aspen is reducing greenhouse gas emissions is by owning hydropower and wind energy sources, and using as little coal as possible to produce electricity. Right now Aspen gets 83 percent of its electricity from renewable wind and hydropower.

approximately 500 local governments. These are local governments and local government associations that are seeking sustainable solutions to their issues and problems. From this beginning, ICLEI is creating case studies for other local governments.

The methodology of ICLEI is aimed at achieving locally based sustainable development It has developed an incremental process of creating political awareness of sustainability and the local environment, then making community-specific action plans with tangible measures of success, and applying tools of project evaluation. It is heavily reliant on the good will and environmental transparency of the local government.

ICLEI develops training, materials, and practical practices in applied areas of local governance around sustainable development. One area of sustainability that local governments can affect is through the materials and services they procure, or use. In the discussion of the U.S. municipal ordinances on the precautionary principle, the procurement processes are heavily emphasized. Governments can often go beyond what they require private citizens to do. Because local governments are closer to the problems of environmental degradation and poverty, they emphasize the procurement of sustainable goods and services. Local governments are huge consumers of products and services, and what they choose to procure can also have the effect of moving the private market to offer more sustainable products and services.

ICLEI also helps local governments adapt to shifts in intergovernmental relations brought about by new global and national emphasis on sustainability. Many traditional environmental programs develop and operate at national and state levels. As social policies of sustainable development gain more traction at the local level, these national programs become more engaged at the local level. Wildlife, biodiversity, parks, and water quality programs are examples. Local governments must also deal with complex issues of culture, economics, poverty, and politics in a policy environment that is both transparent and accountable. This is a challenge for them and for all policies of sustainable development.

CHAPTER 4

Controversies

AID: DISASTER RELIEF MODELS AND INFRASTRUCTURE MODELS

Poverty is a brutal context for any disaster. Natural disasters like floods, drought, fires, and earthquakes cause much more damage to human populations when poverty is widespread. When poverty is in the context of a history of colonialism and heavily indebted nations, as in sub-Saharan Africa, the damage to the ability of people to sustain themselves in way that allows future generations to exist is put in extreme jeopardy. Food security decreases rapidly and vulnerable populations suffer greatly, especially rural populations, women, and children. The typical aid response is to provide food, generally to refugee camps. In this way, the most dramatic impacts of any disaster are lessened. From a sustainable development point of view, however, other aid structures also need to be considered. The ability of an impoverished population to become self-sufficient enough to feed themselves, and to engage in world trade, may require more than food deliveries to refugee camps. Using nongovernmental organizations and civil societies to distribute both food and the capacity to grow food is one way to restructure disaster aid. Meeting the millennium development goals in the context of any given disaster is one way disaster aid is restructured to meet the needs of the present without jeopardizing the needs of the future. These goals were developed without the assumption of a disaster, but were designed to be correctable in midstream. ***See also* Volume 3, Chapter 1: Equity and Poverty.**

References

Paarlberg, Robert. 2009. *Starved for Science: How Biotechnology Is Being Kept out of Africa.* Cambridge, MA: Center for the Study of World Religions, Harvard University Press.

Sachs, Jeffery. 2006. *The End of Poverty: Economic Possibilities for Our Time.* New York: Penguin Press.

Smith, Keith. 2004. *Environmental Hazards: Assessing Risks and Reducing Disasters.* Andover, UK; Routledge.

United Nations. 2008. *Delivering on the Global Partnership for Achieving the Millennium Development Goals MDG Gap Task Force Report.* New York: United Nations. Online at www.un.org/esa/policy/mdggap.

CAPITALISM AND SUSTAINABILITY

Capitalism is a term used to describe free market-based economies both at an ideal or theoretical level, and as a current operating global political economy. Sometimes the free market is also called the "private" sector. This implies a lack of control by government, although in actuality, it is the state that allows many businesses to exist. It is one of the most powerful and prevailing value structures in the world today, as well as in recent history. Many feel that it is poised for a period of robust growth, with far-reaching environmental consequences. In terms of sustainability, there are many fertile grounds for controversy because most versions of capitalism run counter to preservation of environment for future generations. As former Vice President and Nobel Prize winner Al Gore noted:

> Capitalism's recent triumph over communism should lead those of us who believe in it to do more than merely indulge in self congratulation. We should instead recognized that the victory of the West—precisely because it means the rest of the world is now more likely to adopt our system—imposes upon us a new and even deeper obligation to address the shortcomings of the capitalist economics as it is now practiced.
>
> The hard truth is that our economic system is partially blind. It 'sees' some things and not others. It carefully measures and keeps track of the value of those things most important to buyers and sellers. . . . But its intricate calculations often completely ignore the value of other things that are harder to buy and sell: fresh water, clean air, the beauty of the mountains, the rich diversity of life in the forest . . . the partial blindness of our current economic system is the single most powerful force behind what seem to be irrational decisions about the global environment.
>
> Al Gore, 1992, pp. 182–83

Capitalism as a basic concept usually incorporates several characteristics. Private producers of goods and services hire workers. They produce the good or service with the sole intent, and a legal responsibility, to make a profit. Any activity that reduces profits is suspect. Economic development becomes a strong value of the people and the government. The free market development of capitalism becomes developed and supported by government. Corporations are creatures of the state in that they are allowed to form according to the laws passed by the government. The basic form of corporate structure allows corporations to invest money without individual liability for the debts they incur. This decreases the risk of many financial transactions and allows for entrepreneurs, or risk takers, to seek new markets and develop new technologies.

Capitalism as a complicated concept involves multinational corporations with budgets bigger than many nations. Sometimes these nations have natural resources, some that could be irreparably damaged and could be necessary for future generations. A large enough corporation multinational corporation in search of profits might disregard the potential harms to a community or even a nation. For example, a large multinational corporation that mines gold in a very dry desert region might disregard the harm to the ecosystem, surrounding community, and even national economy if the profits from mining were sufficient. Gold mining takes water and discharges arsenic in the process. In a very poor country, capitalism exerted by multinationals may prevent fundamental principles of sustainability from being operationalized. The application of the precautionary principle, where the risks of irreversible damage to natural systems is first assessed, is expensive and time consuming.

Even mining activities are engaging in new environmental activities that pave the way for consideration of sustainable alternatives. The Extractive Industries Transparency Initiative is an effort by an industry trade group to encourage governments to disclose the revenue they receive for oil, gas, and mining operations. By disclosing the revenue from all operations, a more accurate assessment of ecological impact can be made. It does not mean the ecological impact will be less, or that these impacts won't accumulate. Mining operations often have large environmental impacts in product, production, and processes. Nonetheless, there is still a growing global demand for their products.

Capitalism as a complicated concept includes materialistic values, a strong public policy emphasis on economic development, and higher rewards for the strongest competitors. Materialistic values are expressed when identity and status become functions of owning and purchasing. Consuming goods and services for the value of consuming, or social status, and without regard to environmental impacts is materialism. As one scholar on growth noted:

> [C]apitalism itself has answered the demands that inspired 19th century socialism . . . But the attainment of these goals has only brought deeper sources of social unease —manipulation by marketers, obsessive materialism, environmental degradation, endemic alienation . . . we have the freedom to consume instead of the freedom to find our place in the world.

> Clive Hamilton, 2004, pp. 112–13

National, state, and local governments around the world develop many public policies around economic development. Although they are not all capitalistic, they all seek to improve the quality of life of at least present-day humans. The role of government and the type of government intervention into the free or private market are greatly affected by values of capitalism. Governments can respond to strong political pressure from legitimate industry groups and carve out special exceptions or subsidies for industries that show promise of high profits under the name of economic development. Governments can also offer transactional security and privacy for business organizations such as multinational corporations, partnerships, subsidiaries, banks, and other lending institutions.

References

Gore, Al. 1992. *Earth in the Balance: Ecology and the Human Spirit.* Boston: Houghton Mifflin.

Hamilton, Clive. 2004. *Growth Fetish.* London: Pluto Press.

Government Intervention in the Free Market under Capitalism

Government intervention in most forms interferes with perceived profits. Any government regulation that provides for government contracts, grants, or subsidies, however, is sought after by industry. Governmental intervention in this manner is considered misleading by environmentalists because this type of government intervention distorts the true environmental impact of the good or service. Government subsidies, along with avenues other than government power and influence, may exacerbate industrial practices that could jeopardize some of the basic principles of sustainability. Sustainability proponents argue that these government subsidies should go to industries that are trying or supporting sustainability.

At the local level, many towns and villages are seeking industry to improve employment as part of their economic development package. In a capitalistic approach to land and private property the local government will seek to streamline regulatory processes designed to protect the environment and public welfare, abate property and utility taxes for a set period of time, and clean up and give away industrial sites. When regulatory requirements are overlooked at the local level, they can also be overlooked at the state and national level. There are many projects in the United States with significant impacts on the environment that do not file any type of environmental impact assessment. These regulatory requirements are often minimal and act to protect the environment. Without them, overdevelopment, pollution, and unsustainable economic development occurs. The problem is that mispricing goods and services by ignoring environmental impacts of production and consumption causes a misuse of the natural resource base. This upsets sustainability proponents because part of sustainability principles is that systems of life on which we depend are kept in as good a state for future generations.

As capitalism and democracy gain world popularity, some aspects of capitalism may be reaching their environmental limitations. Some environmental amenities have no present value because there is no value to future lives. As countries, states, and cities grapple with growing populations wanting to consume more under capitalism and democracy develop public polices, they will also face environmental limitations. This includes the climate changes to be wrought by global warming, natural resource scarcity, and food shortages. Some current predictions have the greatest climate impact on the equator and surrounding tropical regions. Population and food supply rest on a precarious balance in many nations of this region. If drought and desertification continue along with population growth, then environmental limitations will be much more severe for future generations.

In partial response to concern about global warming and capitalism, a new concept about fairness and sustainability is developing. Climate justice is an emerging concept that focuses on the impacts of climate change. Many climate changes will worsen already bad situations in many parts of the world, and in cities in more parts of the world. Heat islands in urban areas will worsen with increased impacts on people and the environment. Air and water quality quickly become intolerable first during parts of the year and then yearlong. Proactive research on climate models; incorporation of indigenous knowledge; and social, economic, and racial inequalities are all emerging aspects of climate justice. Free market exploitation of environmentally degrading practices and events is generally a controversial issue and will be so in the area of climate change.

Materialistic values in a capitalistic free market that does not place a transactional value on natural resources threaten to consume resources and damage ecosystems. In one way of stating it, the goose that lays the golden egg has been killed; that is, the environment that provides so many resources on which life depends was destroyed before knowledge of the damage was politically powerful enough to find expression in government. That is one reason why sustainability proponents advocate for the precautionary principle. The high demand on natural resources of a growing population in a new global economy burgeoning with capitalism and free markets has not been planned, contains many unknown dimensions, and causes many conflicts and controversies around values. Agenda 21 addressed sustainability in agriculture in the context of population growth estimates:

> By the year 2025, 83 percent of the expected global population of 8.5 billion will be living in developing countries. Yet the capacity of available resources and technologies to satisfy the demands of this growing population for food and other agricultural commodities remain uncertain. Agriculture has to meet this challenge, mainly by increasing production on land already in use and by avoiding further encroachment on land that is only marginally suitable for cultivation.
>
> Paragraph 14:1. United Nations Conference on Environment and Development, Agenda 21, U.N. Document A/CONF.151.26 (1992)

Government facilitation of capitalism as a value and a public policy undergirding economic development also works against sustainability because it supports unsustainable but politically powerful industrial development. This is particularly true in agribusiness. The small, self-reliant family farm is a romantic idea from the past in most industrialized countries. Subsidies of tobacco, cotton, milk, rice, and other crops ignore their environmental impact, but respond to political constituencies at the local, state, national, and sometimes international level. Other nations may choose to subsidize the crops of their own powerful political constituencies. All of these types of government that developed economic policies to attract and enable free markets to flourish may come at an environmental price.

Global Capitalism and Its Relationship to Global Environmental Conditions

The planet has seen a vast expansion of economic activity, even more than population growth. Many environmental writers today believe this rapid and unchecked economic expansion is the main cause of environmental degradation. The world economy now operates on a global scale. Environmentalists, sustainability advocates, and others are concerned that a large population on the verge of economic growth, primarily via some type of "capitalism," will increase the rate of environmental degradation to further decrease the potential for sustainability. As capitalism becomes less an economic theory and more a political cause, it causes a failure to recognize all the nonmarket costs of production. One of the main costs that are ignored under capitalism is environmental impacts. These can occur in the use of raw materials; in the industrial processes of production; in the shipment, storage, and use of the product; and in the disposal of the product. These ignored environmental impacts can accumulate and damage ecosystems. They can take the form of air pollution, toxic and hazardous waste sites, and polluted water. Environmental degradation and public health decline are also part of nonmarket costs. Capitalistic political economies, such as the United States, also support many production activities that produce serious and unaccounted environmental effects. Some examples of these activities are mining, logging, and ranching. Some industries are virtually environmentally unregulated and self-report much of the information. Modern environmental advocacy groups and most approaches to environmental public policy are inadequate to the global challenges presented by current versions of capitalism. Capitalism practiced this way has socialized the harms and risks from speculation and exploitation and privatized the gains from these activities to corporations and their shareholders. Several recent citations to this proposition have to do with bailouts of banks and home mortgage lenders. This has led to the accusation that for some governments, "Banks are too big to fail, and homeowners are too small to bail."

Harnessing Economic Forces for Sustainability

Emerging versions of capitalism are developing new strategies for approaching the demands of society for a sustainable community. The free market is considered important for sustainability because of the innovation and technological advancement that is characteristic of emerging businesses and industry. There are several themes to these emerging theories. Most call for greatly increased efficiency from resource use. Increased efficiency can be in the form of reuse, recycle, and reduced use of natural resources, especially those that are non-renewable. Industrial ecology focuses on the elimination of "wastes" in any production cycles, finding other uses for them and sometimes protecting profit. Many see an economy that moves from the production of goods to the production of services as the necessary step toward sustainability. Ecotourism is often a move toward a service-based economy. Service-sector industries have a high hidden environmental footprint in terms of resource usage, uses of energy, contribution to carbon, and waste generation.

Modern Capitalism

The global economy has grown roughly 5 percent a year, the U.S. economy about 3.5 percent. It is a fast rate of growth, primarily fueled by political capitalism and via multinational corporations. In 2019, the world economy could double in size at a 5 percent growth rate.

Why Is Growth Necessary under Capitalism?

Economic growth is a strong value that undergirds capitalism. The function of growth underscores high and fast growth of products. Economic growth is not limited to capitalism. Other political economies also value economic growth, as noted by one scholar.

> Communism aspired to become the universal creed of the twentieth century, but a more flexible and seductive religion succeeded where communism failed: the quest for economic growth. Capitalists, nationalists—indeed almost everyone, communists included—worshiped the same alter because economic growth disguised a multitude of sins. Indonesians and Japanese tolerated endless corruption as long as economic growth lasted. Russians and eastern Europeans put up with clumsy surveillance states. Americans and Brazilians accepted vast social inequalities. Social, moral, and ecological ills were sustained in the interest of economic growth; indeed, adherents to the faith proposed that only more growth could resolve such ills. Economic growth became the indispensable ideology of the state nearly everywhere.
>
> J. R. McNeill, 2000, pp. 334–36

Growth is traditionally measure by gross domestic product (GDP), although this measure has become more controversial as global concern about the environment has increased.

References

Elkins, Paul. 2000. *Economic Growth and Environmental Sustainability.* London, UK: Routledge.

Hamilton, Clive. 2004. *Growth Fetish.* London, UK: Pluto Press.

Lawn, Philipp Andrew. 2000. *Toward Sustainable Development: An Ecological Economics Approach.* Boca Raton, FL: CRC Press.

McNeill, J. R. 2000. *Something New under the Sun: An Environmental History of the Twentieth Century World.* New York: W. W. Norton.

The Economic and Environmental Consequences of Growth

What is the record of economic development, spurred by burgeoning capitalism and rapid globalization? For sustainability advocates of rapid and radical social changes, this is a key question. The answer for many in the world is cumulative environmental degradation and a widening division between rich and poor. Both the United States and the United Kingdom saw the biggest division between the half with capital and the half without capital since records were kept. Many poor nations and nations heavily in debt cannot afford basic medicines, but wealthy nations continue to accrue great wealth.

The Rich Get Richer and the Poor Get Poorer: Consequences for Sustainable Development

The rich in nation states and between nation states got richer, and the poor got poorer between the mid-1970s and the mid-1990s. Single parents and young people were most often on the getting poorer part of the equation. If sustainable development is to protect future generations, then new questions about increasing poverty among young people and single parents are raised.

The poorest 30 percent of the population in 30 industrialized nations received only 5 to 13 percent of the national total income. The richest 30 percent of the population received 55 to 65 percent of total income. The biggest growth in social inequality were found in the United States, Great Britain, and the Netherlands. The main reason for this finding is that the incomes of the rich have risen quickly and in large amounts. Rich people make much more income than possible to spend in any one lifetime without a concerted effort. By contrast, a large proportion of the poor population lives in permanent poverty, creating a distinct underclass even in the richest nations. These underclasses may not be able to expect the same quality of life for their children as rich people expect for their children.

In Germany, 10.2 percent are poor on an annual basis; in the United States the rate is 14.2 percent and in Great Britain, 20 percent. In all countries, those between the ages of 41 and 50 years have the highest levels of income. Nevertheless, this group is also affected by poverty. Social assistance expenditures in individual countries do not primarily benefit the poor. When big business, conservative political organizations, and the political elites complain about the "high level of welfare payments," their aim is to cut unemployment benefits and social assistance. These are payments traditionally received by the poorest and least privileged in society. These arguments are often tinged by racist stereotypes and attitudes about sexual promiscuity and willingness to work.

The consequence for sustainable development of a growing division between rich and poor is a lack of sensitivity to ecosystem limitations of economic growth. Rich people are able to focus capital assets on natural resource exploitation, and poor people are more willing to sacrifice natural resources for survival.

It is usually the poor of industrialized nations that bear the disproportionate environmental burden unrestrained growth of corporate crony capitalistic growth. It is difficult to persuade nations, communities, and people suffering a low quality of life to act in a sustainable manner for the sake of the planet. The record of ignored environmental impacts is just emerging, and most aspects of it are controversial. Multinational industries reliant on natural resources are exploiting the economic condition of nations suffering a low quality of life, especially in the areas of mining and petrochemical production. Environmentalists charge that they are moving fast because the world community is enacting tougher and more enforceable environmental restrictions on their activities. Industry responds that it is simply reacting to market demand. Environmentalists are now much more aware of global environmental actions. The strength of the environmental movement is its shared observations of nature, because in most places in the world the only observers are the people who live there. The strength of industry is its rapid growth and political power. Over half the largest economies in the world are multinational corporations, not nations.

Economic growth has increased the quality of life and negatively impacted the environment. As measured by GDP, the world economy increased by a factor of 14 from 1890 to 1990. In the same period, the global population increased by a factor of 4, water use increased by a factor of 9, sulfur dioxide emissions increased by a factor of 13, energy use increased by a factor of 16, carbon dioxide emissions increased by a factor of 17, and the marine fish catch increased by a factor of 35. In the first century of capitalism, economic growth increased rapidly and environmental impacts continued to be ignored. Of concern to modern-day sustainability advocates are the unknown cumulative, synergistic, and ecological impacts of such rampant growth because these types of impacts could irreversibly affect natural systems of land, air, and water on

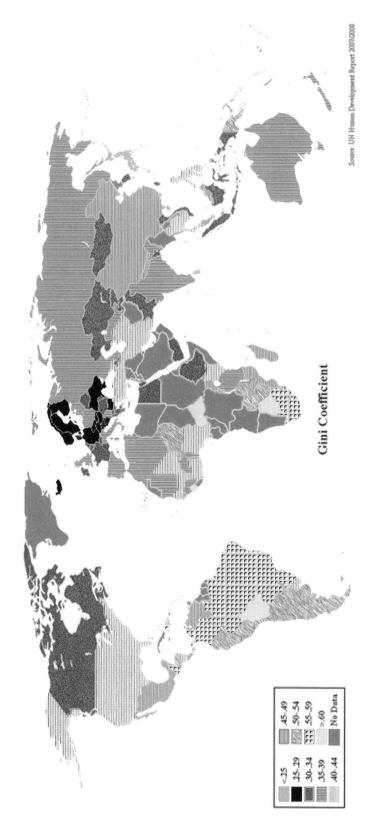

FIGURE 3.6 • World map of Gini coefficient for 2007/2008. The Gini coefficient is a measure of statistical dispersion most prominently used as a measure of inequality of income distribution or inequality of wealth distribution. A low Gini coefficient indicates more equal income or wealth distribution; a high Gini coefficient indicates more unequal distribution. *Source:* UN Human Development Report 2007/2008.

Gini Coefficient

<25
.25–.29
.30–.34
.35–.39
.40–.44
.45–.49
.50–.54
.55–.59
>.60
No Data

which all future life depends. If economic growth continues to increase, the concern is that even more environmental damage could occur.

One dynamic that many feel portends rapid economic growth is the mechanism of global financial markets, which often drive the financing for other markets. High-profit growth is the primary value. One way this is observed is growth in market capitalization and the price paid for its stock. The expected rate of profit growth is important. Losses often cause market value to decrease. In a competitive capitalistic market, most major financing industries need to finance highly profitable industries and corporations to ensure they meet their profit expectations. As finance markets are now global and operate 24 hours a day seven days a week in many places, the ability to quickly and irreparably exploit a weakness in a nation's environmental laws to overuse a natural resource in an unsustainable and profitable manner is much greater. The dynamic of finance markets lead many to believe that some nations could pursue a path of unsustainable economic growth as free markets grow into rapidly increasing populations. All cumulative environmental impacts to date, and what are now small environmental impacts, may increase in scale as population and capitalism increase. The threat of irreversible damage to systems of life like air and fresh water could increase dramatically.

Reference

Medard, Gabel, and Henry Bruner. 2003. *Global Inc.: An Atlas of the Multinational Corporation.* New York: New Press, 2–3.

Recent Evidence of Economic Growth and Environmental Impact

There are many challenges to economic growth as capitalism of any type in many parts of the world. Other nations may have different values and may measure gross national happiness Many locations do not have addresses of personal or commercial property. Land ownership and recordation may be nonexistent in written form. Some value structures do not allow the forfeiture of a home for payment of a debt, perhaps because the property is communally owned, or landownership is unclear. In these types of nations, many lending institutions are hesitant to finance projects because without the security of clear ownership to property as collateral, they could lose profit. Other nations do not have convertible currency, or currency valuable or stable enough to be traded for other currencies on the world market. The relationship of the government to its people is also very different in many places. These three factors—lack of identifiable place, nonconvertible currency, and shifting and unknown nation/people relationships—pose as much of an obstacle for capitalism as they do for sustainability.

A serious challenge to economic growth, and therefore capitalism in many cases, is concern about the environmental impacts. Although the overall trends discussed in the preceding paragraphs describe events over a century, the question becomes more pointed when it addresses whether environmental regulations mitigated environmental impacts enough to be a basis for sustainability. Right now, most researchers answer no. By examining economic growth and environmental impacts since the rise of environmental regulatory regimes, there is economic growth and still environmental impacts, although they are lower than the growth of the world economy. Environmental impacts are still getting worse, but at a slower rate. From 1980 to 2005, according to the World Resources Institute and the Worldwatch Institute, the gross world product increased 46 percent. During the same time period, paper production increased 41 percent, the fish catch increased 41 percent, meat consumption increased 37 percent, energy use increased 23 percent, fossil fuel use increased 20 percent, the world population increased 18 percent, nitrogen oxide emissions increased 18 percent, fresh water use increased 16 percent, carbon dioxide emissions increased 16 percent, fertilizer use increased 10 percent, and sulfur dioxide emissions increased 9 percent. All these measures are fraught with some degree of controversy. They do show that with environmental regulation it is possible to make a difference, but they also show that the difference is not sufficient.

A great concern to sustainability advocates is the characteristic of capitalism to fuel exponential growth. Each year's successive outputs and profits are invested to increase the rate and amount of production. So, too, do waste and environmental degradation exponentially increase. Unlike economic growth that has no theoretical limit, however, the global ecology has definite limits to its capacity to regenerate itself. Once these capacities are ecosystems may reach a tipping point of no return.

Market advocates claim that the market is self-correcting. They point to technology as capable of mitigating environmental impacts because it can reduce the amount of natural resources consumed and produce new products that affect the environment less.

Right now, this is not happening fast enough for sustainability advocates. The latest research from five large West European and U.S. research centers examined these very issues. They concluded:

> Industrial economies are becoming more efficient in their use of materials, but waste generation continues to increase. . . . Even as a decoupling between economic growth and resource throughput occurred on a per capita and per unit GDP basis, overall resource use and waste flows into the environment continued to grow. We found no evidence of an absolute reduction in resource throughput. One half to three quarters of annual resource inputs to industrial economies are returned to the environment as wastes within a year.

James Gustave Speth, 2008, p. 59

References

Hamilton, Clive. 2004. *Growth Fetish*. London, UK: Pluto Press.

Speth, James Gustave. 2008. *The Bridge at the Edge of the World: Capitalism, the Environment, and Crossing from Crisis to Sustainability*. New Haven: Yale University Press.

World Resources Institute, www.earthtrends.wri.org.
Worldwatch Institute, www.worldwatch.org/node/1066/print.

Classic Capitalism and the Environment: Implications for Sustainability

As political economies embrace both capitalism and democracy, many search for the original basis for capitalistic theories. Given their robust growth, sustainability advocates also examine these works for environmental or ecological content. Some of the foundational works of classical capitalism recognized the threat to the environment. One scholar noted in 1944:

> To allow the market mechanism to be the sole director of the fate of human beings and their natural environment . . . would result in the demolition of society. Nature would be reduced to its elements, neighborhoods destroyed and landscapes defiled, rivers polluted, military safety jeopardized, the power to produce food and raw materials destroyed.

> Karl Polanyi, 1944, 73

A fundamental critique of classical capitalism is the preference for the present over the future. The future is always unknown. The value of science to capitalism is in its ability to predict the future within a given probability and range of error. Sustainability expressly values future lives at least to the extent present lives are valued. Future generations cannot be part of current supply-and-demand trends, or markets. Their number and value are unknown, controversial, and may not be quantifiable. When governments try to begin regulatory standards based on sustainability concerns for future generations, they often face this issue of the value of future lives. Because governments are often regulating corporations and their environmental behavior, and because corporations value profits over unknown future lives, corporations often resist any government regulation of their environmental behavior for purposes of sustainability. For example, one aspect of sustainability is the precautionary principle. Here the environmental impact of a proposed activity is assessed to determine if irreversible damage to a natural system could occur. If so, there is usually an exploration of what mitigative actions could and should be taken. All this environmental assessment takes time and ties up capital in what is perceived as a high-risk financial proposition. The emphasis of the present over the future under capitalism also resists expenditures of time and money that do not contribute to profit.

The question of a future value of a human life is a question that is answered in a different context right now. In legal cases of negligence, the value of a wrongful death is computed by juries. One large problem is how to discount future lives. It is impossible to predict how many future lives, whether human or any given species, will exist. There-fore some discounting of future lives is required by cost-benefit analysis of sustainable development. Issues of risky future outcomes are hard to handle under cost-benefit analysis when their probabilities are un-known and the values of future decision makers are unknown. Another problem is that it is hard to value something that is lost forever, such as the extinction of a species. Another problem for cost-benefit analysis is how to value something that has intrinsic value aside from human values. Many environmental economists do this based on a "willing-ness to pay" value. This value is dependent on knowing the tradeoffs between choices now, and for sustainable development, into the future. So much of the environment is just being discovered that values based on present environmental knowledge may be incomplete.

Reference
Polanyi, Karl. 1944. *The Great Transformation*. Boston: Beacon Press.

CHILDREN'S HEALTH

Few would claim to be opposed to protecting the health of children, but in fact protecting children's health is quite controversial at many levels from dose-response measurements to health care. Safe human exposure to chemicals that are commonly encountered where we live, work, go to school, and play are not tested based on the response of children. Children's small body size and developing organ and endocrine systems make them more reactive than adults to exposures to chemicals. Most safe exposures are measured based on the reactions of an adult male. Scientists also tend to exclude women in safety testing because their more complex reproductive systems complicate dose response. This standard clearly disfavors children and women, resulting in greater chemical exposure to these populations, even though they make up the majority of the population.

Children in the City and Sustainability

One major concern of urban residents is the health of their children. The increasing severity of asthma in the United States is concentrated in cities among children who live in poor environments. Air pollution is a major cause of asthma. Children are especially vulnerable to air pollution, compared to adults. Children have more skin surface relative to total body mass. According to the Harlem Lung Center, childhood asthma rates have increased 78 percent between 1980 and 1993. In ad-dition, according to the Columbia Center, there are an estimated 8,400 new cases of childhood cancer each year nationwide. Asthma is a major cause of death among U.S. children, and probably many more across the world. Asthma is also the leading cause of school absences. It is es-timated that about one-third of acute episodes of childhood asthma are environmentally related.

CITIES AND THE ENVIRONMENT

Values play a major role in defining relevant boundaries. Values of industrialization and values of sustainability clash over how to assess the value of land, or real property. As industrial values become postindustrial values, which include some concern for sustainability, the boundary assessment may change to reflect the natural systems of land, animals, plants, water, and air of a given biome. Now, they generally reflect a political subdivision of the state, which is the definition of a municipality in the United States.

Boundary issues are important for ecosystem- or biome-based approaches to sustainability. When they are based on political power rather than natural features, environmental degradation can occur, especially at boundary lines. If the pollution from one country, or city, goes somewhere else, then society has been tolerant of allowing that use. This is the problem of U.S. air pollution causing acid rain that threatened endangered species in the Adirondack Mountains and mountain ecotones in Canada. Property lines are important for the profit-motivated transaction of land, and they translate into municipal boundaries, state boundaries, and national boundaries. They do not translate well when trying to resolve cross-jurisdictional environmental disputes. As pollution and population have increased, so too have the number and ferocity of these disputes. *See also* **Volume 3, Chapter 1: The Urban Context of Equity in the United States.**

Reference

Riddell, Robert. 2004. *Sustainable Urban Planning: Tipping the Balance.* Malden, MA: Blackwell Publishing.

The Exclusion of Cities in U.S. Environmental Policy

In the United States and in most industrialized nations, the land with the most intense ecological footprint, the land that has the weakest enforcement of environmental laws, and the land that will be absolutely necessary for sustainability is decidedly urban. And in the United States, this is one of the most ignored aspects of public and private environmentalism. Most of the pollution is located in cities. In the United States, and in many cities around the world, cities are points of immigration and migration. They have high populations of people of color.

The federal government in the United States began a strong program of environmental regulation around 1970 when the U.S. Environmental Protection Agency (EPA) was formed. Cities were not prioritized, as they were in the formative environmental programs in the United Kingdom. U.S. environmental policy does not require environmental impact assessments for most activities in cities for several reasons. Sometimes the EPA or other federal agency has determined that cities have no complete ecosystems. This means that there could be no significant environmental impacts. According to the National Environmental Policy Act of 1970, if there are no significant impacts on the environment, no full-blown environmental impact analyses need be done. Another reason for the exclusion of U.S. cities from environmental protection is more political. Congress, and sometimes federal agencies, has carved out "categorical exclusions" from the environmental impact programs. For example, a primary source of revenue flow from the federal government to cities comes from the Community Development Black Grant Programs. These funds can be used for many projects, some of which have significant environmental impacts. This grant program was categorically excluded from environmental impact requirements by law. Over the years, many exceptions to environmental impact statement requirements have developed. Environmental impact statements are advisory only. This has crippled the ability of developing an environmental baseline for sustainability in U.S. cities. Some states, such as California and Washington, have state versions of environmental impact statements. These, too, have many exceptions and are advisory only.

On the rare occasion federal environmental impact statements are litigated, they can take a racial turn. In one case in Chicago, African American people moving into a neighborhood were considered by the residents to have a significant environmental impact. In 1972, the Chicago Housing Authority proposed scattered sites for its public housing. The residents of public housing were often African American, and the sites were in white communities. A group called the Nucleus of Chicago Homeowners Association sued the Chico Housing Authority to prevent them from locating the public housing in their neighborhoods on the grounds that it violated the requirement under the National Environmental Policy Act that an environmental impact assessment be performed. They alleged that the public housing tenants had more crime, disregarded private property, and had a lower commitment for hard work. This would have a significant and adverse impact on the aesthetic and economic environment. The case went into a six-week trial with expert witnesses and much media attention. The judge eventually ruled that environments, not people, are pollution. People cannot be ruled as pollutants because of racial stereotypes.

Reference

Harris, William. 2009. *African American Community Development*. Unpublished manuscript.

Equity Impact Analysis

There are proposals to develop an Equity Impact Analysis similar to an Environmental Impact Analysis (EIS). These tend to follow the same basic structure as an EIS but with a focus as a tool for local policymaking. A common ground for measuring adverse effects here is federal Executive Order 12898. Adverse effects can be controversial from an environmental perspective. From an equity perspective, they mean the totality of significant individual or cumulative human health or environmental effects, including:

- Interrelated social and economic effects, which may include, but are not limited to bodily impairment, infirmity, illness, or death

- Air, noise, and water pollution and soil contamination

- Destruction or disruption of man-made or natural resources

- Destruction or diminution of aesthetic values

- Destruction or disruption of community cohesion or a community's economic vitality

- Destruction or disruption of the availability of public and private facilities and services

- Gentrification or displacement of persons, businesses, farms, or nonprofit organizations

- Increased traffic congestion, isolation, exclusion, or separation of minority or low-income individuals within a given community or from the broader community

- Denial of, reduction in, or significant delay in the receipt of, benefits of programs, policies, or activities

Equity impact assessments tend to be focused on local actions such as land use. In terms of an equity analysis, the question of where degrading or risky land uses are allowed and prohibited can affect the adverse impacts the residents experience. Proposed policies with adverse impact potential would be subject to analysis that asks how the policy affects access to jobs, housing, health care, and education. Would the proposed action degrade the quality of life? What would the burdens and benefits be from the policy for each citizen or neighborhood?

Like an environmental impact assessment, an equity impact assessment would define its scope, involve stakeholders, consider draft alternatives, and seek to mitigate impacts. In terms of the equity component of sustainable development, equity impact assessments would be useful because they increase participation, focus on actual land uses, and increase knowledge of ecological impacts.

The cost of excluding cities and their communities from environmental and land equity is a depletion of natural resources. Clean air and clean water are natural resources necessary for a healthy life and they affect regions near and far. Another cost to the equity equation for many city residents and sustainability advocates is that of growing environmental degradation that cannot be escaped by the invocation of privilege or money. Many sustainability advocates fear that current U.S. expensive and time-consuming litigation of antiquated environmental laws creates unnecessary adversaries out of neighbors. This can create a lack of inclusion and fairness when environmental and human concerns are not enough to create a sufficient case of action for judicial intervention.

One primary characteristic of an equity issue is that the consequences of environmental decisions are inescapable for those affected. Although this is a dangerous consequence for socially marginalized people in environmentally marginalized land, with the rise of sustainability it is now dawning on the powerful and privileged parts of U.S. and global society that some environmental privileges are just that—privileges—and they have come at a cost to others. Some argue this is an age-old dynamic of human habitation because someone has to live downstream, meaning there has to be some environmental degradation somewhere. With rising populations, decreasing natural resources, and accumulating pollution pockets, the consequences of environmental decisions are becoming less escapable. Sustainability is the concept that pushes realization of this dynamic for many stakeholders. In some ways, the environment itself is an uncompromising reflection of all our actions.

Urban Ecology

Urban areas hold complex ecosystems in which humans and other animals and plants have coevolved over centuries. As a systematic field of inquiry, ecology often ignored or excluded urban sites. Urban ecosystems have not received the same degree or type of study as those located in less populated areas with some important exceptions such as the Yale School of Forestry Hixon Center for Urban Ecology. Nevertheless, architects and planners of all sorts have experimented with visions for cities and towns that would configure healthful and congenial relationships between humans and their environments.

One primary characteristic of an equity issue is that the consequences of environmental decisions are inescapable for those affected. Although this is a dangerous consequence for socially marginalized people in environmentally marginalized land, with the rise of sustainability it is now dawning on the powerful and privileged parts of U.S. and global society that some environmental privileges are just that—privileges—and they have come at a cost to others. Some argue this is an age-old dynamic of human habitation because someone has to live downstream; that is, environmental degradation must occur somewhere. With rising populations, decreasing natural resources, and accumulating pollution pockets, the consequences of environmental decisions are becoming less escapable. Sustainability is the thought, idea, or concept that pushes realization of this dynamic for many stakeholders. In some ways, the environment itself is an uncompromising reflection of all our actions. As scientists and community residents unravel urban environments with greater knowledge, this reflection will become clearer.

Arcosanti: Merging Architecture and Ecology in the Desert

Arcosanti is an ongoing experiment on the impact of the human-built environment on the immediate ecology. It is located in Arizona, about 65 miles north of Phoenix. It is the brainchild and lifelong passion of architect Paolo Soleri.

Phoenix as a city is known for its sprawling growth and high-impact land use. From 1970 to 1990, the city doubled its size and consumed wilderness at an acre an hour. It is a dry desert environment with seasonal monsoonal weather. Arcosanti stands in stark contrast to such nonsustainable growth and is expressly designed to offer a better alternative. All the building structures are designed to make the most out of sun and shade. Soleri uses many half domes and vaults to move air. The climate of summer and winter determines the shape.

The Cosanti Foundation owns the land, about 860 acres. The environment is high desert, which is cooler than the valley floor. Its elevation is 3,750 feet atop basalt rock outcroppings. It contains some, but not all, community functions in a very compact space. Part of working with an ecological impact constraint is increasing the density of human communities. Arcosanti contains residences of different sizes, woodworking and ceramic shops, a foundry, swimming pool, café, bakery, art gallery, greenhouses, lawns, and a public amphitheater. Everything is to be within walking distance, and little is planned for the car.

Ultimately, Arcosanti is designed to house 5,000–7,000 people in a very compact urban structure. Right now, Arcosanti is still on the grid, but is toying with alternative energy ideas. It is connected to the Internet. Most of the food has to come from the outside. Community areas of health care, education, and religion are lacking. It is only about 5 percent complete. Residents all help build it.

Reference

Antonietta, Iolanda. 2003. *Soleri: Architecture as Human Ecology.* Lima, Peru: Monacelli.

FIGURE 3.7
• Arcosanti, an experimental self-contained town in central Arizona, designed by Paolo Soleri, is seen in July 1980. AP Photo/Suzanne Vlamis.

ECOTOURISM

Tourism is often a significant source of income in developing countries. Tourists bring cash into economies that lack access to cash, and can allow local people to benefit from the natural beauty and uniqueness of their location without destroying it for resource use. Tourism itself, however, has a significant environmental impact, from the greenhouse gases released by traveling to destinations, to freshwater resources tourists use in abundance for washing and other services, to the toll hikers and climbers who may take on sacred sites. Can ecotourism provide a sustainable source of development for communities located in environmentally fragile areas?

The International Ecotourism Society (TIES) defines ecotourism as "responsible travel to natural areas that conserves the environment and improves the well-being of local people." TIES is the oldest and largest organization dedicated to ecotourism around the world. The organization specializes in supporting and advocating for ecotourism development in more than 90 countries and operates under five principles that minimize environmental, cultural, and ecological impacts:

- Build environmental and cultural awareness and respect.

- Provide positive experiences for both visitors and hosts.

- Provide direct financial benefits for conservation.

- Provide financial benefits and empowerment for local people.

- Raise sensitivity to host countries' political, environmental, and social climate.

Ecotourism involves nature-based travel that educates the traveler and is sustainable within the local environment and community. The education and local sustainability elements of ecotourism that distinguish it from traditional tourism. The greatest difference between ecotourism and traditional tourism development, however, is the focus on education, preservation, and conservation of native culture, ecosystems, and biodiversity.

Another important attribute of ecotourism is the engagement of the local community. Research shows that no matter how valuable the natural resources in an area that bring in tourists, if the local communities are not personally benefiting from the increased tourism, the natural resources will not be preserved. Ecotourism also incorporates local communities and helps to develop these areas in a culturally sensitive way. Ecotourism incorporates the environmental conservation, cultural sensitivity, and economic incentives to build sustainable businesses in that community for future ecosystems and future human generations.

In 1992, with the Rio Declaration on Environment and Development, "the Rio Declaration," ecotourism began as a sustainable means of developing a country's tourism market. Many of the principles and goals of ecotourism are incorporated into the Agenda 21 initiatives. Also, in 2000, the Convention on Biological Diversity Guidelines on

Biodiversity and Tourism Development was implemented to help states better understand how and why to protect ecological and animal diversity, and to provide consistent standards globally. The most important recognition so far of how important ecotourism is for the future of the planet, however, occurred when the United Nations named 2002 as the International Year of Ecotourism.

Since 2002, there has been a burst of international ecotourism initiatives, local programs, and certification programs. Some of these include the Environmental Code of Conduct for Tourism, Green Labeling for Tourism, Responsible Travel Guidelines for Africa, Certificate in Sustainable Tourism for Costa Rica, and most recently, The Ecotourism and Sustainability Conference 2008, which addressed greening the tourism industry in the United States and Canada, among other topics.

Ecotourism has created controversy around issues of sustainability. The controversies start with issues of the impact ecotourism has on the environment, on indigenous peoples, and on the local population. There are concerns that ecotourism may decrease the biodiversity of sensitive environments. An increased exposure to people may alone affect the way plants and animals interact in their environment. The increase in waste that often accompanies more tourists may change the behavior of some animals that find it easier to live from wastes than to hunt food. The propensity of tourists to take souvenirs may decrease the actual biodiversity of an area. In some cases, the tourists and the guides may not be fully informed of the environment they are touring and the unintended consequences that may result on a given ecosystem. This is the case in some sensitive tropical rainforests where not all species are yet known. There is some concern that ecotourism may expedite extinctions by direct impacts, and indirectly by introducing invasive species or diseases that eradicate native species.

The relationship between indigenous people and their government may be tenuous. Many governments in areas populated by indigenous people seek economic development through ecotourism. One example of some of the dynamics of ecotourism is in Malawi, a country in Africa. The largest lake there is Lake Malawi. Fishing from these freshwater areas is the main food source for this very poor country. The biodiversity of the fish population is very high. On Lake Malawi is the fishing village of Mdulumanja. In this town is a large tourist hotel. The land, once owned by the country, was sold to a private corporation in 1987. The new owner insisted on tearing the village down to expand the hotel. He then stopped allowing the residents to use his water sources and also stopped buying fish from them. He put up a big fence all around his property. In 1990, after much dispute, he evicted the residents with the help of the national government. More than 70 long-term residents were forced to leave. The hotel owner razed all their homes for tourist development. He expanded his hotel to accommodate the tourists who were mainly white South Africans. The residents lived on the fish from that lake and now had to find other means without access to the lakefront. Ecotourism exploitation is dynamic. Lake Malawi National Park is a world heritage site and may be under the same threat of ecotourism development as the hotel industry marches along the waterfront. There is no monitoring of the impact of this ecotourist industry on the aquatic biodiversity.

The type of job creation under ecotourism is generally in the service sector. Maids, cooks, maintenance workers, drivers, and guides who serve the tourists follow the initial jobs in construction. These are generally low-wage jobs but in places of poverty are considered to be better than nothing. In some places they may exploit cultural and social oppression. The people trained and hired for these jobs may not be from the area. Tourists may not be aware of local cultural dynamics and can unintentionally further political and social oppression. In nations where the indigenous peoples have separate interests, ecotourism may occur as economic development that disrespects native cultures.

Many of the construction practices around ecotourism may not be sustainable, and may have negative environmental impacts on natural systems on which present and future life depends. In places where indigenous peoples have interacted sustainably with their environments in holistic ways for long periods, important environmental knowledge could be lost.

References

Agarwal, Anil, and Sunita Narain. 1991. *Global Warming in an Unequal World: A Case of Environmental Colonialism.* New Delhi, India: Centre for Science and Environment.

Dowies, Mark. 2009. *Conservation Refugees: The Hundred-Year Conflict between Global Conservation and Native Peoples.* Cambridge, MA: MIT Press.

Epler, Megan Wood. 2002. *Ecotourism: Principles, Practices, and Policies.* New York: United Nations Press.

Mowforth, Martin, and Ian Munt. 1998. *Tourism and Sustainability: New Tourism in the Third World.* Andover, UK: Routledge.

ENVIRONMENTAL REPARATIONS

The arguments for reparations to African Americans for the damages done by slavery are well developed. They include the long-term pervasive differences in African American health, income, education, justice, and housing that are traceable to the conditions of slavery and the postslavery era. In addition to these data is evidence that the stigma that attached to race because of slavery also injured the land and the ecology of African American communities. This is the map of environmental injustice first revealed by the Environmental Justice Movement. ***See also* Volume 3, Chapter 2: Environmental Justice.**

Reparations for impacted land and communities are based on a physical and practical rationale that parallels sustainability. Environmental reparations are based on the recognition that heavily burdened systems require repair in order to be able to sustain all forms of life, regardless of the reasons why these places were impacted. Environmental reparations to poisoned lands and watersheds will benefit communities of color and poor communities, as well as revitalizing the living systems on which all living things in those bioregions depend.

References

Brophy, Alfred L. 2006. *Reparations: Pro and Con.* New York: Oxford University Press.

Collin, Robin Morris, and Robert W. Collin. 2005. "Environmental Reparations for Justice and Sustainability." In *The Quest for Environmental Justice Human Rights and the Politics of Pollution*, ed. Robert Bullard. 209–21. San Francisco: Sierra Club Books

Johnston, Barbara Rose, and Susan Slyomovics, eds. 2008. *Waging War, Making Peace: Reparations and Human Rights.* Walnut Creek, CA: Left Coast Press.

LABOR

Labor and Transportation Costs in a Global Market

As businesses search for ways to increase their profits, they closely examine and compare the costs of labor and transportation. The choices that businesses make in these areas dramatically affect communities, the environment, and human health. These two costs can greatly affect the profit made on finished products. In addition, they relate to each other in the global marketplace. Low labor costs in one country may tempt companies to move their manufacturing and service sector businesses to take advantage of these labor savings. These moves often result in a process called "outsourcing" when businesses choose to contract with an outside supplier of labor. Outsourcing of jobs can create hardships for families and communities, sometimes blighting once thriving neighborhoods and cities.

Savings in labor may be lost if the cost of transporting goods to and from low-cost labor locations exceeds the cost savings on labor. As the price of petroleum fuels rises, the cost of transportation may well exceed the cost of labor, making local labor sources attractive, even if they are more expensive. For example, if the cost of labor in one country is much lower than in another country, the incentive to manufacture there is only as great as the relative costs of transporting goods. Any savings on labor costs will be spent on the costs of transportation. Decreasing fuel expenditures may help shrink the carbon footprint of a business. Localization helps to restore local communities.

What makes local sources of labor more expensive is often not simply the hourly wage, but also employer commitments to provide benefits such as health insurance, unemployment insurance, and retirement contributions. These commitments can be voluntarily negotiated by the employer and employees, or mandated by government. Outsourcing labor is a way for employers to avoid these commitments. Expensive transportation costs may exceed the costs of these social benefits, especially if these benefits are subsidized by government in the form of tax breaks to businesses that provide them.

Another way that labor costs can become global issues is through labor immigration. Some business sectors cannot outsource their high labor costs. For example, agriculture and construction labor must be done in a specific place, and laborers must come to those sites. In these sectors, labor savings are often achieved by hiring immigrant labor present in the country on work permits, or completely undocumented. These laborers may be willing to work for less and without benefits because they do not know their rights, do not expect benefits, or will accept less because it is so much more than they can earn in their country of origin. In addition to foregoing wage and benefit protection, these labor forces are also less likely insist on environmental and work safety measures. If they are present illegally, they may fear punishment and deportation more than the environmental and work safety hazards they encounter at work. Therefore, this type of labor force also increases the likelihood of environmental and human health harms. In some areas, and for some crops, the yield is so substantial that it cannot be processed at harvest using only locally available legal laborers.

Market-Based Strategies, Equity, and Sustainability

Market-based strategies to achieve sustainability use conventional trading incentives to achieve goals related to sustainability. Buying and selling objects of trade is at the core of markets. Subsidies are payments that reward conduct deemed desirable. Taxes require payments from individuals or businesses. One use of taxes is as disincentives for conduct that government wants to slow or eliminate.

Some people believe that markets can imitate nature successfully if they close the loops of productivity in ways that imitate natural closed loops. By using various tools of law and government, they argue that the power of the marketplace can reshape our economies to restore natural systems.

Market-based strategies focus on the ability of wealth to dictate outcomes. To the extent that such strategies exclude the poor or those without sufficient of capital to participate in the market, they can re-create the problems they are attempting to solve in a new form. Failure to consider the equity impacts of market-based strategies almost always ensures substantial unintended consequences that undermine the opportunities for success. Inclusion of communities that must live with the consequences is one way to ensure that these consequences are considered at a meaningful stage.

POLITICAL AND ECOLOGICAL BOUNDARIES

The government of a country can exercise power over its territorial areas defined by its physical borders. These borders may not have a relationship to ecosystems or other environmental features. They may reflect other facts of history without regard for the geographical or cultural features of the place. Ecosystems and their features, like watersheds or mountain ranges, often cross borders between countries.

These artificial lines create multiple problems for sustainably managing our resources and communities. First, these lines limit actions in ways that do not translate to effective or efficient ecological intervention. They also limit the communication of important data and information relative to ecosystems and their features. One political response to both of these disconnections is a new emphasis on regionalism in political planning. Any effort to share and distribute power, however, will meet with resistance from those whose power or expectations of power are diminished.

References

Bacon, Christopher M. et al., eds. 2008. *Confronting the Coffee Crisis: Fair Trade, Sustainable Livelihoods and Ecosystems in Mexico and Central America.* Cambridge, MA: MIT Press.

Conca, Ken. 2005. *Governing Water: Contentious Transnational Politics and Global Institution Building.* Cambridge, MA: MIT Press.

Gallagher, Kevin P., and Lyuba Zarsky. 2007. *The Enclave Economy: Foreign Investment and Sustainable Development in Mexico's Silicon Valley.* Cambridge, MA: MIT Press.

Just, Richard E., and Sinaia Netanyahu. 1998. *Conflict and Cooperation on Trans-Boundary Water Resources.* New York: Springer.

Democracy and Sustainability

Basic democratic theory requires the participation and consent of the many that are governed in public law and policy. Exclusion and lack of consent make government action illegitimate. For those who see the need for change toward sustainable behavior as urgent and widespread, the need to consult widely and to develop political constituencies and leadership for sustainable change seems too complex, indirect, and compromised to promise hope of effective change. Some have argued that timely change on the scale that we would need to avoid collapsing our webs of life can be achieved only by a different arrangement of power, such as a plutocracy or oligarchy of properly motivated people or organizations.

One possible response to this argument is that the time taken to ensure genuine participation on a local level ensures compliance and eliminates the need for costly and slow enforcement mechanisms. This is one reason that Agenda 21 recommends that government action toward sustainability be formulated at the lowest effective level of government. Another response lies in the observation that effective action need not mobilize all of the population. Effective changes can be generated by smaller groups that exercise symbolic influence, if not actual power. The power of the tipping point can be exercised voluntarily, even in the absence of governmentally sanctioned power to coerce.

References

Baber, Walter F., and Robert V. Bartlett. 2005. *Deliberative Environmental Politics: Democracy and Ecological Rationality.* Cambridge, MA: MIT Press.

Putman, Robert D. 2004. *Democracies in Flux: The Evolution of Social Capital in Contemporary Society.* London, UK: Oxford University Press.

Svedin, Uno Britt, and Hägerhäll Aniansson. 2002. *Sustainability, Local Democracy, and the Future: The Swedish Model.* Emeryville, CA: Kluwer Academic Publishers.

POPULATION AND CONSUMPTION

There is great debate about what is causing humankind to overuse our ecosystems. Some believe the root cause is overpopulation. Others believe that overconsumption of resources in the developed world contributes far more to the overuse of our resources and environmental degradation. They point to the ecological footprint of developed nations in comparison to developing countries.

Population growth itself is related to human health. As developing countries are able to provide effective public health measures and good nutrition increases, people survive to maturity in greater numbers, which enables greater reproduction. Population also responds to opportunities for women to obtain an education and opportunities to work outside the home. In developed nations, where such opportunities are more available to women, population rates fall over time. In some developed nations, population rates are below replacement levels, which may lead to the need for immigration of young people to maintain adequate work forces and the social benefits that are partly financed by their contributions.

Some basic facts about human population and sustainability are noncontroversial. The human population of the earth is increasing rapidly. The rise in population sometimes is described as a positive feedback loop because gains in population will create more gains. When there are more women of childbearing age, they tend to bear more children. The rates of childbearing women vary widely from country to country. Generally, women in developed nations and wealthier women do not have as many children as women in other nations or socioeconomic classes. Therefore, the population rates in many developed nations are declining while the population rates in other parts of the world are increasing. Overall, the total number of humans on earth continues to rise.

Human population has a variable relationship to resource consumption. Resource consumption varies widely with cultural practices and expectations. Resource consumption in developed countries and among privileged classes is much more intense in amounts and rates of resource use than in other areas. Less developed countries use resources at a dramatically lower rate.

Beyond these facts lie controversies about population control, its significance to sustainability, and what, if anything, to do about it.

Population and the Environment

Some advocates of population control argue that at a certain number of humans on Earth will simply overwhelm the ability of natural systems to function. But that number, as an absolute, is a matter of conjecture. Others argue that at a certain number, the carrying capacity of Earth will be overwhelmed. These arguments are often countered with arguments based on the differences in the rate and intensity of resource and energy use. For example, India and China do not use the resources that Western developed nations use even though their population is larger and growing. If India and China grow in the manner of Western developed nations, creating a comparable environmental footprint, the projected effects on the ecosystems of Earth will be devastating. Under this latter approach, the focus of argument shifts to the mode of development and whether it will be sustainable, not a question of population numbers.

References

Wilkinson, Richard, and Kate Pickett. 2009. *The Spirit Level: Why More Equal Societies Almost Always Do Better.* St. Albans, UK: Allen Lane.

Wilson, Edward O. 2003. *The Future of Life.* New York: Vintage Books.

Zovanyai, Gabor. 2001. *Growth Management for a Sustainable Future: Ecological Sustainability as the New Growth Management for the 21st Century.* Westport, CT: Greenwood Publishing

Two Approaches to Population Control

Human population rates are made up of the numbers of women bearing living children, called a fertility rate, and the rate of death among the population as a whole, together with and immigration over one time period as compared to an earlier or later period. Population can be affected by fertility, death or mortality, and immigration. Human mortality rates are generally decreasing as a function of improved public health, food security, and development. This long-term decrease may be periodically offset by devastating disasters such as drought, floods, epidemics of disease, or famine related to these conditions or war. The focus of population control as a long-term strategy has focused on childbearing and women.

Childbearing is a moral, as well as scientific, issue and many religions and cultures have very specific beliefs and expectations built around this event. Moral and cultural beliefs have shaped two very different approaches to the question of fertility and population control. One approach is focused on the avoidance of pregnancy through abstaining from sexual contact. That is the official approach of the U.S. government both domestically and in foreign affairs. No U.S. foreign aid for any program related to development can be used for any other method of birth control. The same policy applies for educational funding of domestic programs.

A quite different approach is in use by the United Nations. The UN approach has been based on evidence that as women's health, education, and social status improves, women tend to bear fewer children. *See also*, **Volume 3, Chapter 4: The Role of Women.**

POVERTY AND THE ENVIRONMENT

Whether in a developed country, or in a developing country, poverty and the inability to meet basic needs for food, shelter, and care make some human communities even more vulnerable to environmentally degraded conditions. People faced with exposure and hunger will contribute to environmental degradation to meet basic life needs. People driven by insecurity as to basic living conditions are likely to accept employment opportunities regardless of consequences to human and environmental needs.

Commitments to Poverty Reduction

Poverty reduction is seen as necessary to achieve sustainability. As part of this goal the United Nations created the Millennium Development Goals (MDGs). These goals are described as a set of time-bound and measurable goals for combating poverty, hunger, disease, illiteracy, environmental degradation, and discrimination against women. The goals are nation specific and dynamic. There are eight goals with 18 targets to be achieved by 2015 using 1990 as the base year. A total of 48 comprehensive indicators are used to describe whether the targets and goals are met.

Goal one is to eradicate extreme poverty and hunger. The targets within this goal are to decrease by half the number of people who are in extreme poverty from 1990 levels by 2015. According to the UN, the proportion of the population below the national poverty line has decreased from 26.1 in 1990 to 22.7 in 2002. The target to be achieved in 2015 is 13. The incidence of poverty in the seven poorest districts has increased to between 30 and 37 percent. These pockets of poverty are a concern.

Goal seven is to ensure environmental sustainability. The targets within this goal are to integrate the principles of sustainable development within the nation's policies and prevent the loss of environmental resources.

Target ten is to decrease by half the population who does not have access to a sustainably fresh supply of drinking water by 2015, with

1990 as the base year. According to the UN, the proportion of households with access to safe drinking water increased from 68 percent in 1994 to 82 percent in 2001. The target is 86 percent by 2015. For more information see www.adb.org/Statistics/mdg.asp.

In many nations with high rates of poverty, there is a failure of the state to relate to its people. People in many nations with severe poverty live in life-threatening conditions of civil war, starvation, and drought. Leadership in international arenas may not be the same leadership as at the local level. There can be anarchy and failed states. Warlords and tribal chieftains may be the only leaders and may not care about the nation or the international arena of the nation. Environmental exploitation often goes hand in hand with human exploitation. A local custom of tribute to the local governing authority may seem parochial to some, and even corrupt to others. In many places the rule of law extends only by ever-present force and fed by the desperation of severe poverty. In many nations the largest employer is the military, and children facing starvation are easily enlisted. In such situations the challenges of sustainable development are enormous.

Why Poverty Matters for Sustainability

Modern philosophies devoted to protection of the environment have not recognized the dangers that poverty pose for the environment. Human communities suffer as the environment and its supporting webs of life are damaged. The human communities that suffer the most are poor communities. These communities suffer from disruptions in the food cycles and exposures to pollution and waste more than other communities do. Worse yet, many proposals for saving ecosystems increase the burdens on already burdened communities by depriving them of work and other forms of sustenance. In extreme circumstances, these communities will tolerate further degradation in their ecosystems and increased pollution in order to meet the basic needs of life for themselves and their families.

Some former colonial powers have recognized their responsibility for endemic poverty in former colonies. These former colonies provided materials and labor to build great wealth in now-developed nations. They were given large loans of money from international development banks and others to build infrastructure for development, but the money was often not spent on these projects because of fraud and theft. The interest payments on these loans deprive poor countries of money needed for health care, education, and housing. Great Britain has forgiven its former colonies millions of dollars in debt and encouraged other colonial powers and development banks to do the same.

Religious Organizations

Almost all religious organizations recognize some duty, obligation, or recognition of the poor. The Christian religion recently published the

Green Bible. Arising from a religious concern for a dying world, the Green Bible addresses the connections between poverty, environmental degradation, and self-accountability. Some consider the bible as one of the major contributors in writings about caring for the Earth, as a powerful ecological handbook. It addresses awareness, appreciation, stewardship, earth-keeping, environmental fruitfulness, and conservation. The Green Bible has the environmental passages in green ink and refers to other religious texts. The personal connection to the environment is practiced through culture and religion by many communities.

An ecumenical group of spiritual organizations has also joined to advocate for the protection of the environment and provision of assistance to the poor. They drafted the Earth Charter, a statement of principles of responsibilities based on faith. This group is an independent global organization representing more than 2,500 organizations, 400 cities and towns, global agencies like UNESCO (United Nations Education, Social and Cultural Organization) and IUCN, (International Union of Conservation of Nature) and individuals. The charter has four basic principles subdivided into 16 smaller tenets. The four principles are respect and care for the community of life; ecological integrity; social and economic justice, including the eradication of poverty; and democracy, nonviolence, and peace. *See also* **Volume 3, Chapter 1: Poverty; Appendix B: The Millennium Development Goals.**

Saint Francis of Assisi

Saint Francis of Assisi is a saint known for his affinity for the animal and natural world. He wrote the "Canticle of the Creatures," which evidences how intertwined his spirituality and view of nature were. Below is an excerpt as translated by Maurice Francis Egan:

O Most High, Almighty, good Lord God, to thee belong praise, glory, honor, and al blessing.

Praised be my Lord God, with all his creatures, and specially our brother the sun, who brings us the day and who brings us the light; fair is he, and he shines with a *very* great splendor. O Lord, he signifies to us thee!

Praised be my Lord for our sister the moon, and for the stars, the which he has set clear and lovely in heaven.

Praised be my Lord for our brother the wind, and for air and clouds, calms and all weather, by which thou upholdest life in all creatures.

Praised be my Lord for our sister water, who is very serviceable to us, and humble and precious and clean.

Praised be my Lord for our brother fire, through whom thou givest us light in the darkness; and he is bright and pleasant, and very mighty and strong.

Praise be my Lord for our mother the earth, the which doth sustain us and keep us, and bringest forth divers fruits and flowers of many colors, and grass.

References

Chapple, Christopher Key, and Mary Evelyn Tucker, eds. 2000. *Hinduism and Ecology: The Intersection of Earth, Sky, and Water.* Cambridge, MA: Center for the Study of World Religions Harvard University Press.

Folz, Richard C. et al., eds. 2003. *Islam and Ecology: A Bestowed Trust.* Cambridge, MA: Center for the Study of World Religions Harvard University Press.

New Revised Standard Version Bible. 2008. *The Green Bible.* San Francisco: Harper Row.

Privilege

Environmental values that are assumed implicit become explicit when they conflict with each other. This begins to happen when natural resources become depleted and when human populations increase. Groups that have environmental and quality of life privileges that they do not understand as privileges often perceive these privileges as rights. Conflict can arise when these "rights" are threatened. For example, the "right" to clean air is really a privilege. The globalization of environmental quality issues increases the recognition that "rights" guaranteed by a nation are really privileges that a nation cannot guarantee without the express cooperation of the international community. Education about values of sustainability can help these rights become recognized as privileges and thereby decrease conflict.

Among the values in sustainability is intergenerational equity, or an express concern for the future. The incorporation of equity in environmental decision making is recent. Acknowledging through education that environmental quality is more a privilege than a right helps to include all those necessary in the environmental decision-making processes. Even without a full and complete understanding of the contributions of other cultures and species, the environmentally destructive industrial values of the past and the misperceived "rights" of a few may need to be altered if a future with the same rich diversity and life forms is to be preserved. Equity value changes emphasizing the gains made from inclusiveness and diversity will be articulated to promote ecological education for the unknown content of the future. Inclusiveness demands more than the contemporary bilateral discussions between industry and traditional environmental groups; it demands a complete educational process. The dialogue regarding a concept of a sustainable future and how to organize natural resources and allocate them must now include many people who were external to the industrial economic values of environmental privileges. In this manner, some initial dialogues about sustainability will require capacity building and facilitation, which requires time and money.

Sustainability can mean different things depending on a country's economic and social conditions. Environmental privileges may be perceived as basic rights in some nations, and the same is true in the educational area. In some countries, the need to develop is driven by the basic human needs, and in others consumption of resources far exceeds basic human needs. Sustainability may mean reduction in consumption for some, and increasing development without unsustainable methods of production in others. Leadership toward sustainability in educational areas must mutually recognize different approaches for different conditions. This is an example of differentiated leadership in the application of developing sustainable practices.

References

Weisman, Leslie Kanes. 2007. *Discrimination by Design: A Feminist Critique of the Man-Made Environment.* Champaign: University of Illinois Press.

Wildman, Stephanie M. 1996. *Privilege Revealed: How Invisible Preference Undermines America.* New York: New York University Press.

RACE AND WASTE

In the United States, the United Church of Christ researched and published a ground-breaking study in 1987 that demonstrated that the more hazardous the waste, the more likely it was in an African American community. The study showed that it was a 1 in 10,000 chance that this result was random, or a 99.9 degree of certainty. In 2007, researchers revisited this study to see what had changed. The 2007 report indicated that the racial disparities are even greater than first measured as a result of more accurate measures. Geographic information systems, more accurate environmental reporting, and more engaged citizen activism all contribute to a more accurate assessment of environmental impact.

Hazardous wastes are kept in specially designed facilities called treatment, storage, and disposal facilities (TSDFs). The 2007 report found that the proportion of people of color estimated to be within three kilometers of TSDF to be disproportionate. Three kilometers is the distance used because it is commonly accepted as the distance in which adverse health impacts, property values, and quality of life impacts suffer.

The equity component of sustainability includes those most affected by environmental degradation because that knowledge is important for systems of nature to survive for future generations.

Waste has a local, national, and international dimension. As waste gets harder to hide, harder to dispose of, lasts longer, and becomes more hazardous, it gets moved around. Treating waste is generally more expensive than simply moving it to someplace and hiding it. The hiding places can be in a deep mine or in the deep ocean. Oil wastes disposed of by deep well injection in the United States in the 1970s and 1980 in Texas may now be leaching into water aquifers. The Earth is covered with vast oceans and many nations simply dump their wastes into them. There are now islands of plastic wastes floating all over many oceans in the areas where the currents are still. The international waste trade uses poor developing nations as dumpsites for the wastes of developing nations. In all these cases, the environment may suffer in a way that threatens future sustainability. ***See also* Volume 3, Chapter 2: Environmental Justice.**

References

Bullard, Robert D. et al. "Toxic Wastes and Race at Twenty: Why Race Still Matters after All of These Years." *Journal of Environmental Law* 38 (2008):371.

Feagin, Joe R., and Karyn D. McKinney. 2003. *The Many Costs of Racism*. Lanham, MD: Rowman and Littlefield.

RISK PERCEPTION

Related to the science of risk assessment, risk management determines how to plan for and communicate about risks. Risk perception is a science devoted to examining the qualitative aspects of risk, not simply its quantitative aspects. Its main source of controversy lies in the dynamic set of assumptions contained in every risk assessment.

Government often requires a risk assessment to be performed in many areas of environmental and developmental activity. These studies are used to determine funding priorities. Most often, risk assessment is done as a matter of expert assessment. But risk perception is a developing aspect of this science, and a commonly held perception of environmental and ecological risk can provide an important common platform for environmental action that is missing in current controversies. Government has an important stake in developing such common ground as a basis for legislation and regulation for sustainability, especially in changing assumptions of permissions to pollute toward assumptions of eliminating all forms of waste and pollution.

Risk is a way of determining the allocation of resources. Dangers that are real require some plan of action to ensure security. Nevertheless, how much to invest in avoidance depends greatly on the perception of risk associated with that danger, including its probability of occurrence, and the severity of consequences if it occurs. In addition, investment in avoidance varies greatly according to the social and political arrangements underlying contingencies for dealing with danger.

Ecosystem risks connected to rapid and widespread change in our environment and supporting webs of life are difficult to identify and plan for because of their attenuated nature. We do not know enough about the facts of such complicated and interrelated dynamics to fully predict their impact. We do not even know what we do not know. This level of ignorance is called epistemological ignorance. In this situation, it is hard for those enjoying the benefits of their current position to appreciate why there is a need for change. It is also hard for those who are suffering the hardships of their current position to appreciate the need to develop and grow in a different way from the past.

This combination of ignorance, coupled with a contemporary unequal distribution of ecological benefits and burdens, has contributed to a stalemate of political wills in developed and developing countries. Developed countries are faced with the challenge of changing climate and energy resources still enjoy advantages that historical policies conferred.

The need for change of a radical nature still may seem remote. Developing countries faced with a tide of environmental refugees (people whose lives have been disrupted by environmental degradation) and natural disasters are under tremendous pressure to develop resources for the benefit of their suffering populations. The common platform necessary for these countries to agree on an agenda of sustainability is a commonly held perception of the risks of ecological collapses and their interrelatedness. *See also* **Volume 1, Chapter 2: Waste, Pollution, and Toxic Substances.**

THE ROLE OF WOMEN

Many countries include cultural traditions that limit the role that women are allowed to play in their communities. These countries typically do not allow women equal participation in education, employment opportunities, political participation, and other acts of citizenship. The core statements of policy about sustainability call for an end to these practices. Full inclusion of women is extremely controversial. It would require adequate provision of funding for women's education. The idea of women working alongside men earning the same pay is also very controversial. Finally, the principles of full public participation require full facilitation of women's presences and voices in environmental decision making and policy decisions. This will require change in some cultural traditions, some of which are based on religious convictions.

Population control is also a central concern of sustainability. It is also extremely controversial as it relates to sexual activities and the status of women. In addition, contraception is condemned by major religious groups like the Roman Catholic Church, and some forms of contraception such as abortion are illegal in some places. Studies have shown that the greater women's access to education and employment, the fewer children they tend to bear.

Sacred Sites and Sustainability

Sacred sites are places honored by local and indigenous people as places of worship, communion with the spirit world and the dead, and for reasons not disclosed. When such sites are undisclosed by indigenous custom, inclusive decision making may be impossible. Indigenous communities may refuse to make their claims to certain areas known, and refuse the challenges of Western-style proof or evidence of ownership and use.

Some critical decisions about sustainability can impact these areas. Examples include renewable energy development that may be sited on sacred sites or used in the production of energy. Wind farms can sometimes threaten sacred sites. Geothermal phenomenon may be part of sacred ritual. Ecotourism can also affect sacred sites when tourists want to hike or climb terrain that was not trodden by indigenous people.

Ayers Rock in Australia, called Uluru by indigenous peoples there, attracts thousands of tourists each year, but only a few elders were allowed to climb on these rocks by tradition. *See also* **Volume 3, Chapter 4: Ecotourism.**

The Sustainable Assessment: Appraisals and Environmental Assessments

The value of land is dependent on many dynamics: market forces, comparable property, community requirements, and changing environmental values. One emerging dynamic is avoiding cleanup liability and costs. Because of much litigation in the United States, the scope of liability and enforcement of cleanup cost recovery is increasing. This increases financial risk in associated real estate industries like banking and mortgage lending. No one industry wants to be liable for any cleanup costs. This avoidance of cleanup liability often results in no cleanup, or cleanup to very low standards. As waste accumulates and travels via natural systems like water to all parts of a given biome, these sites tend to get worse. This is a large challenge for sustainability advocates. (See Private Property, this volume). Other ways to value land under the description of natural or sometimes social capital are being developed to expand nonmarket concepts of environmental value. Environmentally sustainable concepts do have a market value in many types of construction.

In an unintended consequence, the rush to avoid cleanup liability at all costs has resulted in greater environmental assessment directly at the site. Potential buyers and lenders interview contiguous neighbors, sample soil and water, and search public land and environmental records. They do extensive environmental agency research at the federal, state, and local levels. If the land is found to be contaminated, another level of more probing, on-site assessments takes place. These are not human health, ecological, or cumulative risk assessments that are discussed elsewhere in this volume. They are assessments of cleanup liability and generally assume low levels of cleanup limited only to that site. They may assess whether they can divide the property to avoid the contaminated portion of the site. Generally, at this level an environmental assessment must include the costs of remediation in the assessment. A traditional market value appraisal of real property does not offer much useful environmental information about the land. An environmental assessment, by contrast, includes the actual environmental condition of sites and the cost of cleaning them, without assessment of biological or chemical risks.

UNCERTAINTY IN FACTS AND VALUES

One of the most difficult aspects of reaching decisions in any complex context is the problem of uncertainty about facts. There are different levels of factual uncertainty. Some facts can be known with effort; some cannot be assessed even with effort because of the failure of record keeping or lack of technological skill. Some facts cannot be known because we do not know what we do not know. This is called epistemological uncertainty and it is the most radical level of factual uncertainty. Epistemological uncertainty is characteristic of most ecological crises.

Uncertainty: Accuracy of Fact in the Process

The lack of certainty as to facts and their accuracy is a large part of procedural fairness and has a large role in the expansion of policies of sustainability. If new policies of sustainability have equitable principles, the resultant inclusion will bring more affected stakeholders to the decision-making process. One of the gaping holes of most environmental policy is the lack of fact, or knowledge, of the ecosystem and of the public health risks. Many of the new stakeholders, like urban residents, will not only bring important environmental information, they will bring questions, demands, and policy solutions. As the application of the precautionary principle grows in the United States, this process will make much more stringent demands for accuracy to be fair. The facts or knowledge required for the analysis under the precautionary principle must be enough to answer questions about irreversible impacts on the ecosystem and the people. The intergenerational component of sustainability definitions may expand "people" to mean future generations. Unknown futures, such as the value of future lives may not be possible to know with enough accuracy to answer completely some of the questions required under the precautionary principle. Controversies will arise about whether any new sustainability policy is accurate enough to proceed.

Another aspect of accuracy as part of procedural fairness is the competence of authorities to make factual decisions. There are many small jurisdictions all over the world that make land use decisions of potentially global impact. There are many places in the world where environmental impacts of major global impacts go virtually unregulated, unmonitored, and uncontrolled. Many places simply do not have the capacity to make large environmental decisions quickly that incorporate knowledge about current ecology and potential environmental impacts. In many of these places, one or two stakeholders decide how to make the most profit from natural resources, like mining and logging. As sustainability emerges as a land use decision-making paradigm, the inclusionary aspects will bring in more stakeholders to the decision-making process. Sustainability advocates say this will bring in more facts and increase accuracy to increase the perception of fairness.

The rise in citizen monitoring of environmental decisions is a search for "facts," for more accuracy in the assessment of where they live, work, play, worship, or travel. The procedural aspect of accuracy for fairness requires large amounts of information about the environment and this access is rapidly growing. In the United States, the Toxics Release Inventory lists the emissions, discharges, and wastes of the major industries, and some governments. It is maintained at a federal level and has rapidly developed at the state level. In the United States, right-to-know laws are just starting at the municipal level, and some states are expanding an agricultural right-to-know law. It is likely that much more, and more accurate, environmental information will come into the decision-making process whether it is via sustainability or the need for accuracy in procedural fairness.

CHAPTER 5

Future Directions and Emerging Trends

There are many difficult challenges to implementing concepts of equity and sustainability in the current landscape of environmental decision making all over the world. Many current political and economic realities drive decisions in most societies. The divisions between some groups in some societies are very strong, and in some instances, irreconcilable. This is part of the political reality. This problem is compounded in places where there is little government or a lack of trust of the government. According to some sustainability theorists, this lack of trust in governmental leadership has contributed to its failure to motivate necessary changes in vision, values, and behavior. The failure of leadership can be traced to a failure to broadly embrace all the constituencies that make up popular society and contribute to behavioral and attitudinal realities that make most human habitation unsustainable. This lack of trust of government, and sometimes industry, tends to be greatest among the most oppressed parts of that society and makes their inclusion under government auspices more challenging.

ASSET BUILDING MOVEMENT

The asset building movement in the United States focuses on poor and low-income communities. The movement is designed and advocated as a supplement to the traditional consumption-based welfare programs. The goals of the asset building movement go beyond sustaining the poor in the United States; rather the movement attempts to create structures that would eventually allow the target population to pull themselves permanently out of poverty. Asset-based policies focus on saving for the future and long-term changes in how society addresses the issues of long-term, perpetual poverty. One of the options utilized by the asset building movement is Michael Sherraden's individual development accounts (IDAs). These accounts allow low-income families to save and eventually enter the mainstream financial market. The program operates similarly to an employer match or 401(k) contributions, but instead it matches the monthly savings of working-poor families who are saving toward purchasing an asset. Assets in this situation are considered first homes, paying for post-secondary education, or starting a small business; the diversity of these assets allows each individual or family to rely on its unique strengths to enter the financial mainstream.

The asset building movement is a separate social policy paradigm that supplements, but does not seek to preempt or replace, the current welfare system in the United States. Since its inception, a national asset building movement has successfully worked to increase opportunities for working low-income families and individuals to build necessary financial assets. The movement has been bolstered by increased public awareness of the importance of financial assets and the recognition that merely having a job cannot ensure financial and economic security. Asset building includes a variety of activities connecting low-income residents to financial institutions; financial education; credit cleanup and budgeting; savings opportunities through IDAs and other savings/investment vehicles; microenterprise; earned income tax credit; asset protection approaches such as antipredatory lending measures, retirement planning, and health insurance; and the development of asset building collaboratives.

Mohammed Yunus

Mohammed Yunus is known as the "banker to the poor," also the title of his autobiography. Yunus founded the Grameen Bank in Bangladesh in 1983 after teaching economics at the Chittagong University. While a professor at the university, he took his students on a field-trip to a poor village. He spoke with a woman there who was a basket weaver. She told him about her plight: she needed to borrow money to buy the bamboo for her baskets, and the lender charged weekly interest that took all but a penny profit margin. She was working, but unable to pull herself or her family out of abject poverty. This experience changed Yunus's mind forever. He started to lend small amounts to people personally, mostly to poor women who were in situations similar to that of the woman he met in the village that day. His experience with these small personal loans convinced him that poor people had the desire and ability to participate in businesses to lift themselves from poverty, but they were denied access to capital because there was no way to value their qualities in traditional banking terms.

Against all the advice from banks and friends, he started the Grameen Bank, which focused on microlending as a way to eradicate poverty. Today, this bank has a loan repayment record of 98 percent, with more than 94 percent of borrowers being women. The bank has grown to serve over 2.1 million borrowers and receives $1.5 million a day in weekly installments.

Yunus and Grameen Bank won the Nobel Peace Prize in 2006. In accepting the prize he said, "Lasting peace cannot be achieved unless large population groups find ways in which to break out of poverty. Micro-credit is one such means."

Reference

Karl Grandin, ed. *Les Prix Nobel. The Nobel Prizes 2006.* [Nobel Foundation], Stockholm, 2007 nobelprize.org/nobel_prizes/peace/laureates/2006/yunus-bio.html.
muhammadyunus.org/content/view/47/69/lang,en/.

COLLABORATIVE APPROACHES

Failure to Include Everyone

Another barrier to achieving sustainability may well lie in the inability of present-day leadership and communities to address the current effects of historical inequities. This results from continued exclusion from environmental and land use decisions. These barriers to implementing sustainability are important because to continue to ignore them will increase their effects for already overburdened communities and eventually everyone in those natural systems.

An ecumenical group of spiritual organizations has also joined to advocate for the protection of the environment and provision of assistance to the poor. They drafted the Earth Charter, a statement of principles and responsibilities based on faith. Earth Charter is an independent global organization that represents more than 2,500 organizations, 400 cities and towns, global agencies like UNESCO (United Nations Educational Social, and Cultural Organization) and IUCN (International Union of Conservation of Nature), and individuals. The charter has four basic principles: respect and care for the community of life, ecological integrity, social and economic justice including the eradication of poverty, and democracy, nonviolence, and peace. These principles are further subdivided into 16 smaller tenets.

Collaboration in Environmental Decision Making

Most environmental decision making is no decision at all. Not enough information is known for most environmental decisions. As our population and environmental impacts increase and become more noticeable, however, conflicts and controversies around environmental issues increase. There are several models of environmental decision making. They can be adversarial, as in court where one party sues another. Parties can be citizens, business, and governments. They can use alternative dispute resolution methods that are less adversarial. Both these methods limit the number of people involved, and often do not really resolve the underlying environmental issue in a way that would sustainable. Some stakeholders view the court system as politically hostile to them and may not trust them.

Many questions about the fairness in sustainability are about the benefits and burden of environmental decisions. Many communities want to increase their environmental benefits. Communities that are well organized and have political and economic power can generally deter environmentally degrading land uses. This land use process is called "NIMBY" meaning "not in my backyard." These environmentally degrading land uses are often called "LULUs" or locally unwanted land uses. Communities that are less well organized, or have less political or economic power, are often on the receiving end of a long history of environmental pollution. These communities and individuals want to lessen their environmental exposures and perceive as unfair that they continue to receive disproportionate exposure. This is generally called an environmental justice concern. Many collaboration models arise from environmental justice controversies.

By the time environmental justice issues become recognized as such by the federal government, they tend to:

1. Cut across agency jurisdictions or areas of expertise

2. Involve many stakeholders holding mutually inconsistent perspectives about the nature of the issues confronting them

3. Involve parties having longstanding, adversarial relationships

Since environmental justice communities are the places with high levels of environmental impacts, many environmentally motivated sustainability advocates engage in environmental justice issues seeking collaboration. Much of the U.S. public policy experience with collaboration is with these communities and it developed into an "Environmental Justice Collaborative Model." The elements identified for the environmental justice collaborative model are based on:

1. The experience of all the stakeholders (community, business or industry, and different levels of government)to date with the environmental issues at hand

2. The local ecological and land use history in the community

3. The capacity of all stakeholders to meaningfully engage in collaboration. This refers to the time, commitment, resources, money, and expertise necessary to solve the environmental justice issue at hand. It may also include the need to have stakeholders experienced with collaboration.

Collaborative, multistakeholder environmental decision making requires five basic ingredients to make these three elements meaningful.

1. Getting together: Stakeholders must get together and discuss issues in meaningful ways.

2. Building trust and ownership: Trust between stakeholders is difficult to establish given the history of lawsuits between environmentalists and industry and government. It is also easy to lose, making the next wave a stakeholders engages in mending relationships. Some state processes for environmental justice collaboration, such as New Jersey, may require the stakeholders to be part of any solutions. To be considered for an environmental justice collaborative project in New Jersey, the community had to agree to be part of the solution.

3. Strategic planning: Collaborative processes require goal setting, mission development, and organizational development.

4. Taking action: Collaborative processes are action focused when successful. Because they are so resource intensive, stakeholders expect and work for results.

5. Deepening and broadening the work: Collaboration broadens the range of perspectives on a given issue by including more stakeholders with the capacity to meaningfully engage in decisions they are part of. The project that began as an environmental justice collaborative project often becomes more ecologically based. The broadening of environmental justice collaborative projects to more ecologically based concerns is one of the key junctures of environmental issue and sustainability. This especially touches on the equity component of sustainability because of its focus on public inclusion.

References

Bonorris, Steven, ed. 2007. *Supplemental Environmental Projects: A Fifty State Survey with Model Practices.* Chicago: American Bar Association. (PDF file available at no cost from the Web site of the Environmental Justice Committee of the Section of Individual Rights and Responsibilities, American Bar Association.)

Hemmati, Minu et al. 2002. *Multi-Stakeholder Processes for Governance and Sustainability: Beyond Conflict and Deadlock.* London, UK: Earthscan.

Sabatier, Paul et al., eds. 2005. *Swimming Upstream: Collaborative Approaches to Watershed Management.* Cambridge, MA: MIT Press.

U.S. Environmental Protection Agency: Public Involvement. Available at www.epa.gov/pub licinvolvement/.

COMMUNITY HEALTH MAPPING

Community health mapping (CHM) is a system of building understanding and collaboration between area health care providers and community residents. The goal of CHM is to work together to help people assume more responsibility for their own health and create healthier communities. CHM is particularly helpful for rural communities and encourages a group of people that are often left out of health planning. Health mapping is based on ideas of direct and indirect health care outcomes. CHM can also encourage planning for a community park to encourage leisure-time activity and physical activity, which may in turn help lower the risk of heart disease or other illnesses among the local population.

CHM provides a neutral environment for community participation. It includes participation of community leaders, health care providers, and residents. The plan gives ownership to the community by helping the community plan for its health needs. Not only is CHM a valuable process to empower communities to articulate their own health and well-being issues, but it also has the capacity to challenge the traditional conception of health care and wellness being centered around physicians. The following indicators are used to determine community health:

1. Physical activity

2. Overweight and obesity

3. Tobacco use

4. Substance abuse

5. Responsible sexual behavior

6. Mental health

7. Injury and violence

8. Environmental quality

9. Immunization

10. Access to health care

Asthma and Community Health

In 1999, it was reported that asthma prevalence was 13–20 percent in developed countries, but only 3–5 percent in India and Africa (developing countries). Furthermore, in developing countries the prevalence rate was also different between urban and rural areas; it was almost unheard of in rural regions. In developed countries, however, the rural and urban prevalence of asthma was and is not significantly different. With the asthma epidemic, the prevalence rates for developing countries have significantly increased. Asthma is beginning to affect the people of developing countries in the same way it has in developed countries. Because many people in developing countries do not have access to medical care or medication to make asthma a controllable disease, bronchial asthma is a disease that is becoming a major health issue in many developing countries.

Many factors may have contributed to the rise of the problem of bronchial asthma. Increasing air pollution, fast modernization, and widespread construction work are some of the reasons for asthma to thrive. The situation is complicated by poor access to medical services, high price of effective drugs, and poor health education among the affected population. Increased urbanization may have modified the traditionally low incidence of bronchial asthma in the developing world. Westernization of diets, improvement in standard of living, decrease in exercise rates, increase in the number of dust mites, and more pollution have been blamed.

References

Al-Hajjaj MS. 2008. "Bronchial Asthma in Developing Countries: A Major Social and Economic Burden." *Annals of Thoracic Medicine* [serial online, cited 2008 Jun 23];3: 39–40. Available from: www.thoracicmedicine.org/text.asp?2008/3/2/39/39633.

Hofrichter, Richard, ed. 2000. *Reclaiming the Environmental Debate: The Politics of Health in a Toxic Culture.* Cambridge, MA: MIT Press.

Community Health Indicators

The Community Health Status Indicators project collects data on community health from 3,141 U.S. counties and makes it available online for public viewing in an easy format by county. Data collected include demographics, summary measures of health, national leading causes of death, measures of birth and death, relative health importance, vulnerable popula-

tions, environmental health, preventive services use, risk factors for premature death, and access to care. Data are also displayed for comparison purposes between similar counties in other states. The information may be displayed on maps or downloaded in brochure format. It will be a sparkplug for community action and mobilization around health and environment issues.

GLOBALIZATION AND LOCALIZATION

Globalization of communications facilitated globalization of markets and capital. Globalized markets for goods rely on a transportation network that is dependent on fossil fuels. As these fuels become more expensive and their supply less reliable, the reliance of globalized trade on these transportation networks will change, sometimes dramatically. When the cost of transportation of goods becomes too expensive for consumers to bear, consumers will be forced back into localized goods and products. This is a countermeasure to the effects of globalization moving employment in certain manufacturing and agricultural sectors to distant locations in order to save labor costs. When transportation costs exceed labor costs as a cost of production, localized supplies will regain their market saliency.

Agriculture in particular will need to adapt to the high cost of transportation. Some foresee that this will force localization of foods, that is, increasing the amount of agricultural production near urban areas. This development has been severely compromised by urban sprawl and the environmental impacts of population growth in urban centers taking both land and water resources away from local agriculture. Geographic information systems are used to plot and analyze food deserts in all cities. Not only is agricultural land taken from urban residents, but in sprawling metropolitan areas, basic food sources are often gone. A food desert is a socially distressed part of a city typically with low or no income and no access to healthy and affordable food. A food desert can indicate both areas of public health degradation because of the lack of food and high transportation impacts because of the distance necessary to travel for food.

One idea for localization of food resources in urban areas is vertical farming. Traditional farming is done on the surface of the earth at the level of the topsoil. Among the challenges of traditionally configured farming is the need to acquire and manage large tracts of land usually outside of the market area for which the food is grown. Acquiring large tracts of arable land is expensive and may destroy wetlands or impinge

FIGURE 3.8 • Sky farms or vertical farms grow crops in urban skyscrapers rather than traditional fields that use more horizontal square footage. Illustrator: Jeff Dixon.

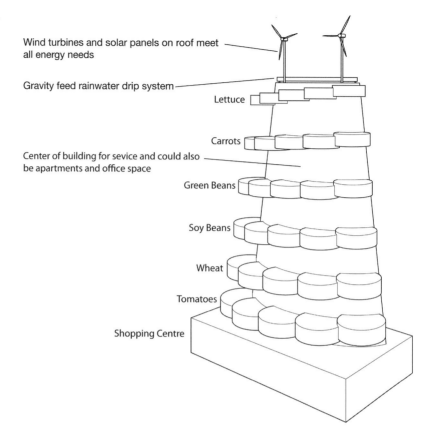

Wind turbines and solar panels on roof meet all energy needs

Gravity feed rainwater drip system

Lettuce

Carrots

Center of building for sevice and could also be apartments and office space

Green Beans

Soy Beans

Wheat

Tomatoes

Shopping Centre

species habitat. In addition, managing such tracts may require substantial investment in machinery and infrastructure such as irrigation or drainage, and farm equipment that uses fossil fuel for its energy. Vertical farming is more similar to greenhouse growing operations, done in multistory buildings resembling skyscrapers. It is compact in its impact on the actual surface acreage, preserving wetlands and other important land uses including species habitat. In addition, locating these sky farm buildings in and near urban markets greatly reduces the cost of transportation as part of the cost of food. *See also* **Volume 3, Chapter 2: Globalization.**

References

Buckingham, Susan, and Kate Theobald. 2003. *Local Environmental Sustainability.* Boca Raton, FL: CRC Publishing.

Hess, David J. 2009. *Localist Movements in a Global Economy: Sustainability, Justice, and Urban Development in the US.* Cambridge, MA: MIT Press.

Riddell, Robert. 2004. *Sustainable Urban Planning: Tipping the Balance.* Malden, MA: Blackwell Publishing.

INDUSTRIAL VALUES

Another barrier to sustainability is the value of development and growth, especially as it applies to natural resources. It may not be possible to renew some natural resources if they are used too much. Restraint of development and growth could upset expectations of private property owners. If decisions about sustainability are made as other environmental decisions are currently made, then certain bioregions and human sacrifice zones will continue to develop. These land areas are simply forsaken because of environmental hazard or have a low market value because they are too polluted to clean up. This happens because of artificially imposed limits in thinking about the nature of bioregions, and because of the institutional exclusion of the residents in political and economic decision making about land, environment, and even voting. It will not be possible continue to exclude residents under sustainability.

MILLENNIUM DEVELOPMENT GOALS

Key indicators to watch for information about our progress toward sustainability in the area of equity and fairness are the eight Millennium Development Goals:

1. End Hunger

2. Universal Primary Education

3. Gender Equity

4. Child Health

5. Maternal Health

6. Combat HIV/AIDS

7. Environmental Sustainability

8. Global Partnership

These eight goals are to be achieved by the year 2015. Each goal has specific targets and indicators that have been identified. The United Nations Development Programme monitors progress toward the achievement of these goals and publishes its findings regularly. *See also* **Appendix B: The Millennium Development Goals.**

PLACE STUDIES

The environmental history and current condition of any one place are important for all sustainability policies and programs. Environmental inequities of the past and present can help explain the current ecological status and help develop meaningful environmental baselines for most communities. For most cities, we do not yet have the environmental information necessary to establish environmental baselines. Without environmental baseline measures, it is difficult to decide what types of behavior changes to develop, what types of rules to enforce, and how to begin a sustainable program. This is especially true in the area of cumulative impacts. *See also* **Volume 3, Chapter 2: Environmental Justice.**

References

Buckingham, Susan, and Kate Theobald. 2003. *Local Environmental Sustainability.* Boca Raton, FL: CRC Publishing.
Jacobs, Jane. 1961. *The Death and Life of Great American Cities.* New York: Random House.

Lack of Identification with Place

Another core barrier in contemporary society is a lack of identification with place. This is often described as a lack of a sense of community. Places where people relate to and support each other intergenerationally are said to have a strong community, with many affiliated ties in social institutions like religion, education, neighborhoods, and shared experience with the environment. Ethnocentrism, or a biased perception of other's culture, often prevents an identification with a loss of community other than one's own. Groups that have experienced limited mobility, such as African Americans in the United States, may identify strongly with a particular neighborhood. Limited transportation mobility occurred because of racial discrimination in access to trains, planes, and cars. People know each other better in a neighborhood that lacks mobility because of the greater role of mutual assistance. One of the factors that can contribute to the organic ability of most indigenous people to express a concern for future generations and for the integrity of their environment lies in their traditional sense of identification with place over a period of generations extending both backward in the past and forward into the future. Contrast this with contemporary industrial society and its policies of encouraging rapid and continuous movement of labor forces around the planet to get the cheapest cost for labor. It is a challenge for sustainability to conduct inclusionary dialogues that express concern for place and for future generations across splintered ethnicities and a changing community base.

Even for long-term community members, inadequate notice about important land use and environmental decisions prevent active engagement and inclusion. Added to this barrier of ethnocentrism of place is the whole political economy of the land itself. Ownership of private property and control of the land are directly related to socioeconomic and political power, with no consideration to the necessities or carrying capacity of a bioregion.

References

Cannavo, Peter F. 2007. *The Working Landscape: Founding, Preservation, and the Politics of Place.* Cambridge, MA: MIT Press.
Fuchs, D. A. 2003. *An Institutional Basis for Environmental Stewardship: The Structure and Quality of Property Rights.* New York: Springer.

Mapping: Geographic Information Systems

Computer-based technology is now available to show many data sets distributed spatially and oriented geographically. This technique of mapping uses geographic information systems (GIS). Maps are powerful ways to communicate information to audiences across language barriers, educational and informational differences, and other barriers. Mapping technology is widely available; some GIS applications are available without charge on the Internet through companies like Google.

For example, at scorecard.org information about pollution is organized by zip code and searchable by communities in that fashion (www. scorecard.org). Community health statistics are also available and mapable by county through the U.S. Department of Health and Human Services (www.communityhealth.hhs.gov). These approaches to the presentation of data are effective tools for educating communities about important features of community sustainability.

SCIENCE FOR POLICY

Public Policy Analysis and Sustainable Development

Public policies are those policies and practices developed by the chief executive or executive group of government, legislation, law, court decisions, and administrative agencies. The government pursues different policies through rules, regulations, taxes, and subsidies. These polices have the force of law and are the means by which sustainability advocates pursue social change. All kinds of government have public policies including monarchies, democracies, republics, and colonies. The main intent of all these policies is generally to protect the health, safety, order, and welfare of society. Many policies are also designed to facilitate business and trade. Sometimes governments are not representative of all the people and develop policies to protect select constituencies. If these constituencies create degrading environmental impacts, then confrontation and change arise between these groups and sustainable development advocates. At other times, the state or nation and its government do not reach all the people or ecosystems in a given geographic region. For example, in many parts of the equatorial rainforests, there are regions virtually ungoverned by any nation, but all geographic regions are claimed by one nation or another. An example is the many competing claims to parts of the North Pole, especially now that more land mass there is being revealed as a result of the effects of climate change. Public policies about sustainable development in these regions of known biodiversity and rapid and controversial deforestation face challenges when there is no governmental public policy that protects natural systems on which all life depends.

Public policies and personal practices need to change in many places to begin the process of sustainable development. To the extent public policies around sustainable development affect personal patterns of consumption, they can be effective in reducing environmental impacts.

Public policy is by its nature political. Groups that are excluded from politics are often excluded from public policy, even those policies designed for them. Environmental values, power struggles, and controversy mark early sustainable development policy. Public policies determine the distribution of benefits and burdens that affect human behavior. Policy issues of regulatory efficiency, how costs and benefits are measured, the participation and access of stakeholders in the process, and the distributional impacts of the policy all become necessary for policies of sustainable development. All these policy issues are also dependent on values, and sustainable development is dependent on meaningfully incorporating environmental values. Most policies are evaluated by the indicators of value in that society. Environmental values are new to public policy to the extent needed for sustainable development. In the United States and most industrialized nations, economic values are indicators of value. Only those costs that are measurable are emphasized. Intangible costs of cumulative impacts on public health and on ecological integrity are generally not counted because they are too intangible. This is problematical under sustainable development because not only do the effects on the environment accumulate, they can do so without predictable impacts on the environment. As one scholar notes:

> Whereas some impacts pass away soon after their termination, leaving few if any marks of their occurrence, the effects of other activities may be cumulative. This is obviously so in the case of certain physical events but may also be characteristic of more purely sociocultural impacts . . .
>
> The effect of accumulation may not always conform to the expectations derived from simple arithmetic. For example, the impact of one offshore support vessel working out of a small harbor might be slight; the impact of a second might double the impact but still be slight. After some increase in the number of vessels, however, qualitative changes may commence. New docks, new fuel delivery routes, changes in the proportions of persons employed in the hydrocarbon versus fishing industries and their consequent places in the local economies political arena could result.
>
> Roy A. Rappaport, 1994, 167

Pain, suffering, the beauty of nature, the costs to future generations, and the value of irreplaceable natural systems are also intangible and generally not captured by traditional public policy analysis. Freedom is also highly valued, not captured by traditional cost-benefit analyses, and transcends governments. As noted by Alexis de Tocqueville in *Democracy in America:*

> Freedom has appeared in the world at different times and under various forms; it has not been exclusively bound to any social condition, and it is not confined to democracies. Freedom cannot, therefore, form the distinguishing characteristic of the democratizing ages. The peculiar and preponderating fact which marks those ages as its own is the equality of conditions; the ruling passion of men in those periods is the love of this equality.

The treatment of all people equally resonates with equity aspects of sustainability and with sustainability generally. Unequal treatment of people in the implementation of young environmental regulatory policies results in pockets of waste accumulation. The unequal distribution of environmental benefits and burdens sparked the U.S. environmental justice movement. In the United States, "freedom" is a strong value often associated with ownership of private property. Individual property owners, including large industrial land users, are theoretically "free" to use their land, water, and air any way they want as long as there is no direct danger to the owner or to others and their private property. The problem develops when the land uses result in environmentally degrading impacts that threaten public health or ecological well-being. As the quotation observes, however, it is the love of equality of all people that is most characteristic of freedom, not the ownership of land. Public policies of sustainable development and private property at local, state, and federal levels, and both ecoregionally and globally face strong stresses as the evidence of our impact on global natural systems forces a reexamination and reprioritization of values. An area that sees many of these stresses first is the area of public policy analyses.

Public policy analyses tend to be based in economics only. They measure economic impacts and related behavior changes of some public policies. Many public policies are political, serving to meet the needs or acquiesce to the demands of large constituencies or stakeholders. For example, many natural resources owned by the U.S. federal government are leased to private corporations for a variety of environmentally degrading uses such as grazing, logging, and mining. In what many sustainable development advocates describe as perverse, some very damaging environmental activities are subsidized by the government. For example, some governments provide subsidies to mining operations, which leave the earth very damaged, or to agribusinesses, which exhaust the soil like tobacco. Governments provide these subsidies to facilitate a particular goal like energy independence or food security. In doing so, however, they fail to consider impacts to the environment and community that may be long term or more diffuse in nature. In terms of environmental policy development, the foundational economic value often misses or undervalues ecosystem services or environmental impacts. Public policy analysis that relies on economic growth as a foundational value also often misses or undervalues ecosystem services or environmental impacts.

Nonetheless, sustainable development will be subject to traditional public policy analyses. Whether these models can be adapted to longer term and more ecologically based sustainable development is unknown. From an ecologically based sustainable development perspective, the answer often depends on the value of the environment, the value of getting information about ecological impacts, and the scale of time. If an action or project threatens to destroy systems on which all life depends, then sustainable development would require a precautionary investigation into ecological impacts. Can traditional public policy analyses capture these aspects of sustainable development?

Reference

Rappaport, Roy A. 1994. "Human Environment and the Notion of Impact." In *Who Pays the Price? The Sociocultural Context of Environmental Crisis,* ed., Barbara Rose Johnston, 167. Washington, DC: Island Press.

Cost-Benefit Analysis

Cost-benefit analyses are often used to set priorities within and between public policies. The main goal of this analysis is to develop a public policy that most efficiently uses resources to solve the given public policy problem. Cost-benefit analysis places a monetary value on the consequences of a given public policy. It can be applied to projects and policies. Both costs and benefits must be expressed, or valued, in monetary terms. Many economists call this "getting the prices right" for environmental or ecosystem services. It is a very controversial analysis in environmental circles. Advocates of true or full cost accounting believe that these benefits can be expressed in quantifiable terms, and they insist that doing so will change our economic calculations and actions toward sustainability.

Under sustainable development, many aspects of nature are priceless, or at least outside the realm of any one generation paying for it. As noted by scholars:

> The basic problem with narrow economic analysis of health and environmental protection is that human life, health, and nature cannot be described meaningfully in monetary terms: they are priceless . . .
>
> There is no reason to think that the right answers will emerge from the strange process of assigning dollar values to human life, human health, and nature itself, and then crunching the numbers. . . . formal cost benefit analysis often hurts more than it helps.
>
> But for most people, there are matters of rights and principles that are beyond economic calculation. Setting boundaries of the market helps to define who we are, how we want to live, and what we believe in. There are many activities that are not allowed at any price.

Frank Ackerman and Lisa Heinzerling, 2004, 8–9

Another area of cost-benefit analysis is human health risk assessment in environmental policy making. Cost-benefit analyses are seldom applied to ecological risk assessments, which are needed for sustainable development. In terms of human health risk assessment, the application of cost-benefit analyses results in the application of the concept of marginal utility to human life and environmental protection. Marginal utility means that the costs have outweighed the benefits of a given activity. When applied to human risk assessment, it means that the cost of removing the last 0–50 percent of a given pollutant may be more than the value assigned to the human lives affected by the chemical release.

A market-based public policy tends to accept the marginal utility of human life. In a private sector economy, it is difficult to "force" a private business to take a loss for the advancement of social policy. Many communities, environmentalists, and others find allowing any marginal utility an affront to the value of human life. Regulatory policies of sustainable development need to take this into account domestically and globally. Some economists, such as Lawrence Summers, have advocated for moving heavy polluting industries into the poorest, least regulated nations under marginal utility concepts.

References

Ackerman, Frank, and Lisa Heinzerling. 2004. *Priceless: On Knowing the Price of Everything and the Value of Nothing.* New York: New Press.

Sagoff, Mark. 1988. *The Economy of the Earth: Philosophy, Law, and the Environment.* New York: Cambridge University Press.

Cost-Utility Analysis

One method used in U.S. health care policy is cost–utility analysis. This is effectively a modification of the cost-effectiveness analysis. The modification is designed to focus on the quality of life issues around health care. The main measure of outcomes is a "quality adjusted life year" (QALY), which measures the number of years of life gained by a health policy adjusted for the quality of the extra years of life.

Cost-utility analyses are used when the quality of life is the main outcome of the policy, when comparing morbidity and mortality, and when comparing cost-utility analyses with other similar analyses.

Quality of life measures are important for sustainable development. Humans are parts of ecosystems and biomes, and degradation of ecosystem health usually results in degradation of human health.

Cost-Effectiveness Analysis

A more modern tool of public policy analyses is the cost effectiveness analysis (CEA). CEA is a technique for comparing the relative value of various policies. In its most common form, a new policy is compared with current policy that is sometimes called the "low-cost alternative." Although there are variations depending on agency and their mission, most use the following process. Each one of these stages in the process can become quite involved. First, the public policy problem is framed. At this stage a number of policy options are considered and brainstormed. Next, appropriate outcome measures for the success of the policy are developed. The next stage in the process identifies the best public policy choice and the costs of that outcome. Sometimes the next stages of these processes develop decision trees to analyze public policy alternatives.

If a policy is "cost-effective," it means that the new policy is a good economic value. Cost-effective does not mean that the public policy actually saves money or achieves program goals. Cost-effective public policy analyses require a value judgment. CEAs used in U.S. public policy is a specific type of economic analyses in which all costs are related to a single, common effect, or public policy goal. It allows decision makers to compare different policies in similar terms and scales. It will evaluate public policy options quantitatively within the confines of a defined model. It requires measurable costs and outcomes, and tends to ignore intangible measures. As the accompanying sidebar demonstrates, some early sustainable development land use policies are applying CEAs to early results.

Sustainability Impact Assessment: Tools for Environmental, Social, and Economic Effects of Multifunctional Land Use in European Regions

Sustainability of land use in European regions is a central point of policy and management decisions at different levels of governance. Implementation of European policies designed to promote and protect multifunctional land use requires the urgent development of robust tools for the assessment of the impacts of different scenarios on environmental and socioeconomic sustainability. SENSOR is the name of a tool used to facilitate sustainable land use. The technical objective of SENSOR is to build, validate, and implement sustainability impact assessment tools (SIAT), including databases and spatial reference frameworks for the analysis of land and human resources in the context of agricultural, regional, and environmental policies. The scientific challenge is to establish relationships between different environmental and socioeconomic processes as characterized by indicators considered to be quantitative measures of sustainability. Scenario techniques will be used within an integrated modeling framework, reflecting various aspects of multifunctionality and their interactions. The focus will be on European-sensitive regions, particularly those in accession countries, as accession poses significant questions for policymakers regarding the socioeconomic and environmental effects of existing and proposed land use policies. SIAT will use the statistical and spatial data continuously collected by European and regional agencies. SENSOR will deliver novel solutions for integrated modeling, spatial and temporal scaling and aggregation of data, selection of indicators, database management, analysis and prediction of trends, education, and implementation. SIAT will be made available to decision makers at the European and regional level, providing user-friendly interfaces and scientifically sound procedures for the assessment of environmental and monetary responses of policy options.

Risk Analysis—Shared Risk: Cumulative, Synergistic Measurement of Risk

Risk assessment is a form of analysis of the probability and magnitude of harm from various events and activities. It is widely used to make decisions. Insurance relies on this computation of risk of harm in making decisions about whether to insure, and if so, how much to charge for insurance. Risk assessment is also used in decisions about development projects, and it is used by governments in budgets and planning activities.

Related to the science of risk assessment, risk management determines how to plan for and communicate about risks. Risk perception is a science devoted to examining the qualitative aspects of risk, not simply its quantitative aspects. Government often requires a risk assessment to be performed in many areas of environmental and developmental activity. These studies are used to set priorities and determine funding priorities. Most often, risk assessment is done as a matter of expert assessment. Risk perception, however, is a developing aspect of this science, and a commonly held perception of environmental and ecological risk can provide an important common platform for environmental action that is missing in current controversies. Government has an important stake in developing such common ground as a basis for legislation and regulation for sustainability, especially in changing assumptions of permissions to pollute toward assumptions of eliminating all forms of waste and pollution. *See also* **Volume 3, Chapter 1: Human Health and the Environment.**

Reference

Whiteside, Kerry H. 2006. *Precautionary Politics: Principle and Practice in Confronting Environmental Risk.* Cambridge, MA: MIT Press.

Ecological Risk Assessment

Ecological risk assessment studies how the land, air, and water interact in a given ecotone or bioregion. These assessments have been performed in sensitive environments such as mountains and deserts where there is little human disturbance or interaction. Their application to the areas that would be included under the inclusionary aspect of sustainability is just starting in the United States. With the application of the precautionary principle to environmental decision making, where it will be necessary to determine if there are any threats to natural systems on which all life depends, ecological risk assessments will be necessary. Because of their newness, time, and labor requirements, ecological risk assessments could be one of the most time-consuming aspects of sustainability decision making.

Exactly how does an ecological risk assessment work? It depends on the place being studied. Some areas are quite complex, such as urban coastlines. Other areas are simpler in terms of lower biodiversity such as mountaintops. Applying the theory of ecosystem risk assessment to real life-and-death situations is a difficult policy challenge. The advancement of ecologically based environmental policy is just beginning and still stumbles on bitter, gridlocked controversies in science, politics, and values. For purposes of sustainability, however, it is important to move forward with ecological risk assessments. Although baseline data for most areas are sorely missing from records and perhaps current knowledge, many industrialized countries such as the United States and Japan have begun to clean up industrialized areas for reuse. In the United States, cleanup policies currently focus on Superfund cleanups. They use ecological risk assessment to determine cleanup methods, mitigation strategies, and cleanup levels. The ecological risk assessment theory applied by the EPA describes, and sometimes predicts, what happens to a bird, fish, plant, or other nonhuman organism when it is exposed to a stressor, such as a pollutant. In the U.S. EPA science-based approach, an ecological risk assessment "evaluates the likelihood that adverse ecological effects may occur or are occurring as a result of exposure to one or more stressors." An ecological risk does not exist unless an exposure has the ability to cause an adverse effect. Adverse impacts are narrowly construed under current ecological risk assessments methods, with much attendant controversy. Further, that exposure must contact an ecological component to "cause" the identified adverse effect.

EPA Ecological Risk Assessment: Groundwork for a Sustainable Future?

As social pressure increases for more sustainable approaches to environmental decision making, the ecological risk assessment process will be called into play. Many are quick to point out its many faults but, it is a U.S. starting point for beginning dialogues around cities and ecology. According to the relevant law, Superfund's goal is to:

PRINCIPLE 1: REDUCE ECOLOGICAL RISKS TO LEVELS THAT WILL RESULT IN THE RECOVERY AND MAINTENANCE OF HEALTHY LOCAL POPULATIONS AND COMMUNITIES OF BIOTA

PRINCIPLE 2: COORDINATE WITH FEDERAL, TRIBAL, AND STATE NATURAL RESOURCE TRUSTEES

Coordination with all nearby interests is awkward and new in U.S. environmental policy because of the litigation potential and the adversarial nature of those proceedings. Stakeholders such as nearby property owners are protecting their legal interests in property. *Coordinate* is a term of art that means giving notice of all potentially important environmental action at a minimum. The notice should be timely and accessible. Coordination can mean much more, too. It can mean facilitating meetings and collaborating on solutions to environmental problems. Under equitable aspects of sustainable decision making, this aspect of coordination could be significantly expanded.

PRINCIPLE 3: USE SITE-SPECIFIC ECOLOGICAL RISK DATA TO SUPPORT CLEANUP DECISIONS

According to the cleanup regulations, site-specific data should be collected and used to determine whether or not site releases present unacceptable risks and to develop quantitative cleanup levels that are protective of people and the environment. The use of site-specific data is a big step toward sustainability. It is difficult to make generalizations that are accurate from models of one ecosystem or biome. It is it often much more expensive, time consuming, intrusive, and controversial to get site-specific data than it is to make generalizations from models. Under the sustainability precaution principle, site specific-data are required to make the judgment of whether a proposed activity irreversibly damages natural systems necessary for life in that biome.

PRINCIPLE 4: CHARACTERIZE SITE RISKS

This is a traditional part of most assessments of risk: human health risk assessments, comparative risk assessments, and cumulative risk assessments. When evaluating ecological risks for the potential for biome preservation, the EPA ecological risk assessments examines risk in terms of:

1. Magnitude, or how large the risk is

2. Severity, or how disruptive of natural systems or public health

3. Distribution of the risks—are people or natural systems being disproportionately impacted?

4. The potential for recovery of the affected natural systems such as water

PRINCIPLE 5: COMMUNICATE RISKS TO THE PUBLIC

The Superfund law requires EPA ecological risk assessors to communicate to the public the scientific basis and ecological relevance of the assessment. Communication of risks to the public is far from public involvement or community environmental empowerment.

Limited and late versions of ecological risk communication characterize most current environmental decision making. Under principles of inclusion and precaution within sustainability, much better communication to a more diverse group of stakeholders is necessary.

PRINCIPLE 6: REMEDIATE UNACCEPTABLE ECOLOGICAL RISKS

Superfund's express policy goal is to eliminate unacceptable ecological risks. Contaminated natural systems that affect the ability of local populations of plants or animals to recover and maintain healthy populations at or near the site should be cleaned to a level that protects these local populations. This is an important building block for future policies of sustainability.

There is controversy, however, about what an acceptable level is. An acceptable level of cleanup to a community with robust participation may be different from those communities without environmental leadership. If the area under consideration is a biome, then what is acceptable in terms of sustainability will be based in ecological risk assessment processes. In the United States, acceptable levels can vary depending on land use. If the cleanup is industrially zoned land, then a lower level of cleanup is acceptable, which is much less expensive. This partially cleaned natural resource land, however, remains in the biome for the community, Communities and sustainability advocates want sites cleaned up to public health standards.

THE EPA APPROACH: ENOUGH FOR SUSTAINABILITY?

The term *cumulative risk assessment* covers a wide variety of methodologies and approaches: economic, psychological, sociological, and scientific. The U.S. EPA has used risk assessments in much of its growth as an environmental regulatory agency, but has stopped short of cumulative risk assessments. EPA risk assessments are basic, human health risk assessments that describe and quantify the risks of adverse health and ecological effects from synthetic chemicals, radiation, and biological stressors.

Is this type of risk assessment enough to plan for sustainability? A critique of the application of this type of risk assessment to new needs is that populations and vulnerable groups are omitted or ignored. To many environmentalists and community residents, it seems that EPA risk assessments are designed to facilitate economic development, as opposed to protect and preserve the environment. These assessments do not capture the range of public health and ecological assessment necessary for cumulative risk assessments. There is substantial dispute over exactly what is an "adverse impact."

The U.S. EPA is a relatively new federal agency formed in 1970. Early risk assessments were designed to distribute, but not prevent, environmental risks, which were narrowly construed. Ecosystem and ecological risk assessment emerged from human health-based risk assessment and comparative risk assessment. The EPA is just beginning to use some of its first versions of an ecological risk assessment in the cleanup of large polluted sites. Ecological risk assessments are most often conducted by the EPA during the remedial investigation/feasibility phase of a cleanup response process. They are used to evaluate the adverse ecological effects that occur because of exposure to physical or chemical stressors. Adverse impacts is a controversial term meaning, in this type of ecological risk assessment, any physical, chemical, or biological entities that can induce adverse responses at a given site.

The EPA published Ecological Risk Assessment and Risk Management Principles for Superfund Sites in1999 to begin policy implementation. This pioneering and sorely needed policy describes six principles for EPA to consider when making ecological risk management decisions. The EPA is wedded to conservative scientific principles but is moving forward with ecological risk management actions. It wants to be able to present them to the public and other stakeholders. This is a foundational step toward sustainability, especially the renewed emphasis for robust public participation and access to real-time, site-specific environmental information. The actual number of sites engaged with current ecosystem risk assessment is very small. Many of the pollutants at these sites migrate from the site to damage natural systems like water. Almost all rudimentary ecosystem risk assessments to date assume the goal is to prevent ecological degradation, which is also a principle of sustainability.

Superfund Cleanups: Foundation for Ecological Risks in Sustainability Assessments

Superfund is a federal EPA policy and law that mandates the cleanup of some polluted sites. The 1,200 or so worst sites are placed on a National Priorities List. The liability for cleanup of these polluted sites is complex and litigated frequently. According to Superfund law, the producers of the waste, then the shippers, then the storage providers, and ultimately the owner of the property are liable for the cost of the cleanup. These parties are known as primary responsible parties (PRPs). If the PRPs do not clean up the site, then the EPA can do so and go after the PRPs for the cost of the cleanup and put liens on the property until the costs are paid. Private property owners can be liable for waste they did not cause by simply owning the land. To be culpable and liable for an environmental cleanup with no more than ownership of property is a strong policy step in the right direction for sustainability. PRPs can sue other PRPs for costs. This complex method of litigating cleanups has not been particularly effective, but is under constant revision.

Cumulative Environmental Impacts

Cumulative Risk

The by-products of rapid and recently regulated industrialization are wastes and chemicals. Most urban areas in the world have more than 100 years of industrial accumulation in addition to human and animal wastes and by-products. These by-products of industry and community have many labels and names: discharges, de minimus emissions, fugitive emissions, human impacts, environmental or ecosystem impacts, and risks. No matter what the label, they all contribute to cumulative impacts on the environment and sometimes in people. When chemicals accumulate in people, it is called bioaccumulation. Some chemicals bioaccumulate and are detected more readily than others, such as heavy metals.

Cities are particularly vulnerable to accumulating environmental impacts because of their density. When inclusionary dialogue under equitable aspects of sustainability include urban residents, cumulative environmental and ecological impacts will probably pose the most threat to natural systems on which all life depends, a key part of the definition of sustainability. They will also be a concern of the new participants who can possess unique knowledge about the environmental history of a place.

Few environmental policies truly embrace control of cumulative environment impacts. Many environmental policies are relatively new, and many aspects of risk assessment are controversial. A core problem is that no one wants to be responsible for the prior waste of another. This has serious business and economic consequences because no one industry wants to be responsible for the pollution of others, whether from unknown past or current competitors.

CUMULATIVE EFFECTS

Cumulative risk assessment combines risks. Which risks are examined and how risks are combined can differ greatly. Simplistic cumulative risk assessments add up risks for a total risk number or range of probabilities. To be counted as a risk, the activity generally must cause an adverse effect, the sum total of these effects being the cumulative effects. As with most risk assessments, a reaction must be scientifically proven to be caused by the activity. From there it must be medically established that it caused an "adverse" effect. More sophisticated cumulative risk assessments examine multiple sources of environmental stress, examine them to see how they specifically combine, and measure effects based on real populations, not models of a population. Cumulative risk assessment also looks at nonchemical environmental stressors. Most risk assessments focus on the movement of a single chemical or chemical compound through one part of the ecosystem, such as the air or water.

These controversies around risk assessment all point to cumulative risk assessment as part of a continuous environmental system as the best basis for a risk-based approach to sustainability. The EPA has strong interest in sustainability and its policy on cumulative risks is always evolving. To date, the EPA still resists regularly incorporating cumulative risk analysis in most applied policy areas such as environmental impact statements. It may be that international requirements for trade will push the U.S. EPA to begin to implement a policy that includes cumulative risk assessment.

The aggregation of risks is generally based on an assumption of chemical additivity by the EPA. This adds the risk per chemical for a sum total of risk. It ignores chemical antagonism, when chemicals mitigate the risk from one another. In the European Union, synergized risk and risk to vulnerable populations determine public health evaluations. The U.S. approach does not generally do this, and when combined with awkward definitions of "adverse," it becomes less than comparable with more comprehensive assessments of risk. U.S. approaches to pollution control leave many pollution sources completely unregulated, and those that are regulated emit millions of pounds of chemicals per year.

Environmentalists want all chemicals regulated, or at least monitored, from all sources. As environmentalists move into the inclusionary dialogues of sustainability, they will find a good forum to push this issue forward. For an accurate cumulative risk assessment, all past and present environmental impacts must be counted.

How Can Cumulative Effects Be Measured?

There are approximately three primary methods for analyzing cumulative effects.

1. The carrying capacity analysis method identifies thresholds (as constraints on development) and provides mechanisms to monitor the incremental use of unused capacity. Carrying capacity in the ecological context is defined as the threshold of stress below which populations and ecosystem functions can be sustained. In the social context, the carrying capacity of a region is measured by the level of services (including ecological services) needed by the community. The strengths of this method are that it is a true measure of cumulative effects against a threshold, it addresses the effects in a system context, and it addresses time factors. Its weaknesses are that it is currently difficult to measure this kind of capacity directly, there may be multiple thresholds, and this type of regional information in the United States often is not developed.

2. Ecosystem analysis explicitly addresses biodiversity and sustainability. It uses natural boundaries (e.g., watersheds) and applies ecological indicators. Ecosystem analysis entails a broad regional perspective. Its strengths are that it uses regional scale and addresses a large range of ecological interactions (synergy, antagonism, catalysis), addresses time, and seeks sustainability. Its current weaknesses are that it is limited to natural systems, requires more data than we currently have or require, and some of the landscape indicators are still under development.

3. Social impact analysis addresses cumulative effects related to sustainability of human communities by focusing on variables such as demographics, community and institutional structures, and political, social, and economic resources. It projects future effects using social analysis techniques such as linear trend projections. Its strengths are that it addresses social issues and the models provide definitive, qualified results. Its weaknesses are that utility and accuracy of results are dependent on data quality and model assumptions, and social values are highly variable over time.

An engaged community with capacity could select a combination of the preceding factors that would tailor a community assessment to its place and its people. These methods would capture unintended consequences and impacts of urban environmental policy interventions.

Back to the Future: Havasupai

Sustainable communities may be old or new. Some of the oldest sustainable communities exist among indigenous people who often had a long-term relationship with the environment and a concern for future generations.

The Havasupai people are indigenous people of the Grand Canyon. They have lived there for at least 700 years. One sustainable practice of the Havasupai was their use of cottonwood trees, which they tended almost as a crop. Approximately every three years, they would strip the cottonwood trees of their limbs. This did not damage the tree and sometimes encouraged even more growth. The limbs were used as fence posts to enclose animals. Cottonwood is a fibrous, moist wood, but by using the branches as fence posts, the Havasupai would dry them out so that they could become firewood after about three years. Some of the limbs would sprout and be allowed to grow into new cottonwood trees.

Food from the environment was locally attained and completely used. The piñon

FIGURE 3.9
• Havasu Falls, before the dam break of August 2008. ©Juan Fernandez III, M.D. Used by permission.

tree was used for fuel, tar, building materials, and scent, and the nuts were a staple food source. Juniper bushes were also used for building, heat, and food. Rabbits, deer, and porcupine were wintertime food, and all parts of the animal were used. When the Havasupai were denied access to their winter ranges outside the canyon by the U.S. government, their lives became very hard. They brought their fight for their original homelands to Congress in the early 1970s. It was a bitter struggle that took years. On January 4, 1975, President Ford signed P.L. 93–620. This was the largest amount of land the U.S. government ever returned to a tribe. Once the Havasupai had their land restituted, they commenced to reviving old sustainable practices.

Ecopsychology

Human survival has often meant buffering and separating human life from the forces and patterns of the environment and ecology of the place inhabited even as we manage them for our benefit. In developed societies, this separation often exists at a high level reflected in not knowing or thinking about the source of our food, the cost of our lifestyles, the amount of waste we generate, and where it goes. This separation and disconnection from the natural forces and cycles are also thought by some to be responsible for human psychic and mental conditions ranging from anxiety and depression to more dangerous and dysfunctional conditions. A branch of psychology and therapy has developed this idea into a practice of treating humans by reconstructing connections between individual humans and the ecology of the places in which they live.

Reconnection with the natural forces that govern our well-being on Earth is essential to knowing and safeguarding those forces so that they may continue to benefit lives that depend on them. Destructive and dangerous adaptations occurring from climate change, acidification of the oceans, and loss of species are clearly the result of a failure to understand human relationships within and to the ecosystem providers on which all life depends.

References

Roszak, Theodore, Mary E. Gomes, and Allen D. Kanner, eds. 1995. *Ecopsychology: Restoring the Earth, Healing the Mind*. San Francisco: Sierra Club Books.

Wilson, Edward O. 1984. *Biophilia*. Cambridge, MA: Harvard University Press.

Appendix A: Portal Web Sites

Aarhus Convention Implementation Guide at: www.unece.org/env/pp/acig.htm

Agenda 21, www.un.org/esa/sustdev/documents/agenda21/index.htm

Community Benefit Agreements, www.goodjobsfirst.org/pdf/cba2005final.pdf

Environmental Justice Collaborative Model: A Framework to Ensure Local Problem-Solving (EPA300-R-02–001, February 2002, www.epa.gov/compliance/environmental

Executive Order, Federal Actions to Address Environmental Justice in Minority Populations and Low-Income Populations, www.epa.gov/Region2/ej/exec_order_12898.pdf

Good Neighbor Agreements, www.cpn.org/topics/environment/goodneighbor.html

International Council for Local Environmental Initiatives, www.iclei.org/

Majora Carter, 6/8; Interview: 11:22 a.m. EDT, Fri June 6, 2008, www.cnn.com/2008/TECH/06/05/carterinterview/

Mixx Digg Facebook del.icio.us reddit StumbleUpon MySpace

Majora Carter, www.ssbx.org/MajoraCarterStaffBio.htm

Muhammed Yunnus biography, nobelprize.org/nobel_prizes/peace/laureates/2006/yunus-bio.html

Muhammad Yunnus, muhammadyunus.org/content/view/47/69/lang,en/

Model Guide To Public Participation, Www.Epa.Gov/Compliance/Resources/Publications/Ej/Model_Public_Part_Plan.Pdf

Neal, Ruth and April Allen. Environmental Justice: An Annotated Bibliography. www.ejrc.cau.edu/annbib.html

Nelson Mandela, 2006, The Ambassador of Conscience Award, Amnesty International, available online at www.amnesty.org

Principles of Environmental Justice, www.ejnet.org/ej/principles.html

Towards an Environmental Justice Collaborative Model: An Evaluation of the Use of Partnerships to Address Environmental Justice Issues in Communities (EPA/100-R-03–001 and EPA/100-R-03–002, January 2003), www.epa.gov/evaluate

Unintended Impacts of Redevelopment and Revitalization Efforts in Five Environmental Justice Communities, p.3 (online at http://epa.gov/compliance/resources/publications/ej/nejac/redev-revital-recomm-9–27–06.pdf)

UN Division for the Advancement of Women, www.un.org/womenwatch/daw/

UN Environmental Programme, www.unep.org

UN Global Compact, www.unglobalcompact.org/index.html

UN, General Civil Society Hearings, www.un.org/ga/civilsocietyhearings/

UN, Human Settlements Programme, Habitat Agenda, ww2.unhabitat.org/hd/hdv10n2/4.asp

UN, Human Settlements Programme, ww2.unhabitat.org/

UN, Non-Governmental Liaison Service, www.un-ngls.org/

UN, World Urbanization Prospects, esa.un.org/unup

US Government Accountability Office, Environmental Right-to-know, www.goa.gov/new.items/d08115.pdf

Van Jones Home and Bio at www.vanjones.net/page.php?pageid=3, andcontentid=29

Van Jones, advocate for social justice and shared green prosperity. By David Roberts. Mar 20, 2007 www.grist.org/news/maindish/2007/03/20/vanjones/

World Bank, Social Capital

Appendix B: Millennium Development Goals

Official list of MDG indicators

All indicators should be disaggregated by sex and urban/rural as far as possible.

Effective 15 January 2008

Millennium Development Goals (MDGs)

Goals and Targets

(from the Millennium Declaration) Indicators for monitoring progress

Goal 1: Eradicate extreme poverty and hunger

Target 1.A: Halve, between 1990 and 2015, the proportion of people whose income is less than one dollar a day

1.1 Proportion of population below $1 (PPP) per day
1.2 Poverty gap ratio
1.3 Share of poorest quintile in national consumption

Target 1.B: Achieve full and productive employment and decent work for all, including women and young people

1.4 Growth rate of GDP per person employed
1.5 Employment-to-population ratio
1.6 Proportion of employed people living below $1 (PPP) per day
1.7 Proportion of own-account and contributing family workers in total employment

Target 1.C: Halve, between 1990 and 2015, the proportion of people who suffer from hunger

1.8 Prevalence of underweight children under five years of age
1.9 Proportion of population below minimum level of dietary energy consumption

Goal 2: Achieve universal primary education

Target 2.A: Ensure that, by 2015, children everywhere, boys and girls alike, will be able to complete a full course of primary schooling

2.1 Net enrolment ratio in primary education
2.2 Proportion of pupils starting grade 1 who reach last grade of primary
2.3 Literacy rate of 15–24 year-olds, women and men

Goal 3: Promote gender equality and empower women

Target 3.A: Eliminate gender disparity in primary and secondary education, preferably by 2005, and in all levels of education no later than 2015

3.1 Ratios of girls to boys in primary, secondary and tertiary education
3.2 Share of women in wage employment in the nonagricultural sector
3.3 Proportion of seats held by women in national parliament

Goal 4: Reduce child mortality

Target 4.A: Reduce by two-thirds, between 1990 and 2015, the under-five mortality rate

4.1 Under-five mortality rate
4.2 Infant mortality rate
4.3 Proportion of 1-year-old children immunised against measles

Goal 5: Improve maternal health

Target 5.A: Reduce by three quarters, between 1990 and 2015, the maternal mortality ratio

5.1 Maternal mortality ratio
5.2 Proportion of births attended by skilled health personnel

Target 5.B: Achieve, by 2015, universal access to reproductive health

5.3 Contraceptive prevalence rate
5.4 Adolescent birth rate
5.5 Antenatal care coverage (at least one visit and at least four visits)
5.6 Unmet need for family planning

Goal 6: Combat HIV/AIDS, malaria, and other diseases

Target 6.A: Have halted by 2015 and begun to reverse the spread of HIV/AIDS

6.1 HIV prevalence among population aged 15–24 years
6.2 Condom use at last high-risk sex
6.3 Proportion of population aged 15–24 years with comprehensive correct knowledge of HIV/AIDS
6.4 Ratio of school attendance of orphans to school attendance of nonorphans aged 10–14 years

Target 6.B: Achieve, by 2010, universal access to treatment for HIV/AIDS for all those who need it

    6.5 Proportion of population with advanced HIV infection with access to antiretroviral drugs

Target 6.C: Have halted by 2015 and begun to reverse the incidence of malaria and other major diseases

    6.6 Incidence and death rates associated with malaria

    6.7 Proportion of children under 5 sleeping under insecticide-treated bednets

    6.8 Proportion of children under 5 with fever who are treated with appropriate anti-malarial drugs

    6.9 Incidence, prevalence and death rates associated with tuberculosis

    6.10 Proportion of tuberculosis cases detected and cured under directly observed treatment short course

Goal 7: Ensure environmental sustainability

Target 7.A: Integrate the principles of sustainable development into country policies and programmes and reverse the loss of environmental resources

Target 7.B: Reduce biodiversity loss, achieving, by 2010, a significant reduction in the rate of loss

    7.1 Proportion of land area covered by forest

    7.2 CO_2 emissions, total, per capita and per \$1 GDP (PPP)

    7.3 Consumption of ozone-depleting substances

    7.4 Proportion of fish stocks within safe biological limits

    7.5 Proportion of total water resources used

    7.6 Proportion of terrestrial and marine areas protected

    7.7 Proportion of species threatened with extinction

Target 7.C: Halve, by 2015, the proportion of people without sustainable access to safe drinking water and basic sanitation

    7.8 Proportion of population using an improved drinking water source

    7.9 Proportion of population using an improved sanitation facility

Target 7.D: By 2020, to have achieved a significant improvement in the lives of at least 100 million slum dwellers

    7.10 Proportion of urban population living in slums

Goal 8: Develop a global partnership for development

Target 8.A: Develop further an open, rule-based, predictable, nondiscriminatory trading and financial system. Includes a commitment to good governance, development and poverty reduction—both nationally and internationally

Target 8.B: Address the special needs of the least developed countries. Includes: tariff and quota-free access for the least developed countries' exports; enhanced programme of debt relief for heavily indebted poor countries (HIPC) and cancellation of official bilateral debt; and more generous ODA for countries committed to poverty reduction

Target 8.C: Address the special needs of landlocked developing countries and small island developing States (through the Programme of Action for the Sustainable Development of Small Island Developing States and the outcome of the twenty-second special session of the General Assembly)

Target 8.D: Deal comprehensively with the debt problems of developing countries through national and international measures in order to make debt sustainable in the long term

Some of the indicators listed below are monitored separately for the least developed countries (LDCs), Africa, landlocked developing countries and small island developing States.

Official development assistance (ODA)

8.1 Net ODA, total and to the least developed countries, as percentage of OECD/DAC donors' gross national income

8.2 Proportion of total bilateral, sector-allocable ODA of OECD/DAC donors to basic social services (basic education, primary health care, nutrition, safe water and sanitation)

8.3 Proportion of bilateral official development assistance of OECD/DAC donors that is untied

8.4 ODA received in landlocked developing countries as a proportion of their gross national incomes

8.5 ODA received in small island developing States as a proportion of their gross national incomes Market access

8.6 Proportion of total developed country imports (by value and excluding arms) from developing countries and least developed countries, admitted free of duty

8.7 Average tariffs imposed by developed countries on agricultural products and textiles and clothing from developing countries

8.8 Agricultural support estimate for OECD countries as a percentage of their gross domestic product

8.9 Proportion of ODA provided to help build trade capacity Debt sustainability

8.10 Total number of countries that have reached their HIPC decision points and number that have reached their HIPC completion points(cumulative)

8.11 Debt relief committed under HIPC and MDRI Initiatives

8.12 Debt service as a percentage of exports of goods and services

Target 8.E: In cooperation with pharmaceutical companies, provide access to affordable essential drugs in developing countries

8.13 Proportion of population with access to affordable essential drugs on a sustainable basis

Target 8.F: In cooperation with the private sector, make available the benefits of new technologies, especially information and communications

8.14 Telephone lines per 100 population
8.15 Cellular subscribers per 100 population
8.16 Internet users per 100 population

Appendix C: The Earth Charter

United Nations Earth Charter

PREAMBLE

We stand at a critical moment in Earth's history, a time when humanity must choose its future. As the world becomes increasingly interdependent and fragile, the future at once holds great peril and great promise. To move forward we must recognize that in the midst of a magnificent diversity of cultures and life forms we are one human family and one Earth community with a common destiny. We must join together to bring forth a sustainable global society founded on respect for nature, universal human rights, economic justice, and a culture of peace. Towards this end, it is imperative that we, the peoples of Earth, declare our responsibility to one another, to the greater community of life, and to future generations.

Earth, Our Home

Humanity is part of a vast evolving universe. Earth, our home, is alive with a unique community of life. The forces of nature make existence a demanding and uncertain adventure, but Earth has provided the conditions essential to life's evolution. The resilience of the community of life and the well-being of humanity depend upon preserving a healthy biosphere with all its ecological systems, a rich variety of plants and animals, fertile soils, pure waters, and clean air. The global environment with its finite resources is a common concern of all peoples. The protection of Earth's vitality, diversity, and beauty is a sacred trust.

The Global Situation

The dominant patterns of production and consumption are causing environmental devastation, the depletion of resources, and a massive extinction of species. Communities are being undermined. The benefits of development are not shared equitably and the gap between rich and poor is widening. Injustice, poverty, ignorance, and violent conflict are widespread and the cause of great suffering. An unprecedented rise in human population has overburdened ecological and social systems. The foundations of global security are threatened. These trends are perilous—but not inevitable.

The Challenges Ahead

The choice is ours: form a global partnership to care for Earth and one another or risk the destruction of ourselves and the diversity of life. Fundamental changes are needed in our values, institutions, and ways of living. We must realize that when basic needs have been met, human development is primarily about being more, not having more. We have the knowledge and technology to provide for all and to reduce our impacts on the environment. The emergence of a global civil society is creating new opportunities to build a democratic and humane world. Our environmental, economic, political, social, and spiritual challenges are interconnected, and together we can forge inclusive solutions.

Universal Responsibility

To realize these aspirations, we must decide to live with a sense of universal responsibility, identifying ourselves with the whole Earth community as well as our local communities. We are at once citizens of different nations and of one world in which the local and global are linked. Everyone shares responsibility for the present and future well-being of the human family and the larger living world. The spirit of human solidarity and kinship with all life is strengthened when we live with reverence for the mystery of being, gratitude for the gift of life, and humility regarding the human place in nature.

We urgently need a shared vision of basic values to provide an ethical foundation for the emerging world community. Therefore, together in hope we affirm the following interdependent principles for a sustainable way of life as a common standard by which the conduct of all individuals, organizations, businesses, governments, and transnational institutions is to be guided and assessed.

Principles

I. RESPECT AND CARE FOR THE COMMUNITY OF LIFE

    1. Respect Earth and life in all its diversity.

        a. Recognize that all beings are interdependent and every form of life has value regardless of its worth to human beings.

        b. Affirm faith in the inherent dignity of all human beings and in the intellectual, artistic, ethical, and spiritual potential of humanity.

2. Care for the community of life with understanding, compassion, and love.

   a. Accept that with the right to own, manage, and use natural resources comes the duty to prevent environmental harm and to protect the rights of people.

   b. Affirm that with increased freedom, knowledge, and power comes increased responsibility to promote the common good.

3. Build democratic societies that are just, participatory, sustainable, and peaceful.

   a. Ensure that communities at all levels guarantee human rights and fundamental freedoms and provide everyone an opportunity to realize his or her full potential.

   b. Promote social and economic justice, enabling all to achieve a secure and meaningful livelihood that is ecologically responsible.

4. Secure Earth's bounty and beauty for present and future generations.

   a. Recognize that the freedom of action of each generation is qualified by the needs of future generations.

   b. Transmit to future generations values, traditions, and institutions that support the long-term flourishing of Earth's human and ecological communities.

In order to fulfill these four broad commitments, it is necessary to:

II. ECOLOGICAL INTEGRITY

5. Protect and restore the integrity of Earth's ecological systems, with special concern for biological diversity and the natural processes that sustain life.

   a. Adopt at all levels sustainable development plans and regulations that make environmental conservation and rehabilitation integral to all development initiatives.

   b. Establish and safeguard viable nature and biosphere reserves, including wild lands and marine areas, to protect Earth's life support systems, maintain biodiversity, and preserve our natural heritage.

   c. Promote the recovery of endangered species and ecosystems.

   d. Control and eradicate non-native or genetically modified organisms harmful to native species and the environment, and prevent introduction of such harmful organisms.

   e. Manage the use of renewable resources such as water, soil, forest products, and marine life in ways that do not exceed rates of regeneration and that protect the health of ecosystems.

    f. Manage the extraction and use of nonrenewable resources such as minerals and fossil fuels in ways that minimize depletion and cause no serious environmental damage.

6. Prevent harm as the best method of environmental protection and, when knowledge is limited, apply a precautionary approach.

    a. Take action to avoid the possibility of serious or irreversible environmental harm even when scientific knowledge is incomplete or inconclusive.

    b. Place the burden of proof on those who argue that a proposed activity will not cause significant harm, and make the responsible parties liable for environmental harm.

    c. Ensure that decision making addresses the cumulative, long-term, indirect, long distance, and global consequences of human activities.

    d. Prevent pollution of any part of the environment and allow no buildup of radioactive, toxic, or other hazardous substances.

    e. Avoid military activities damaging to the environment.

7. Adopt patterns of production, consumption, and reproduction that safeguard Earth's regenerative capacities, human rights, and community well-being.

    a. Reduce, reuse, and recycle the materials used in production and consumption systems, and ensure that residual waste can be assimilated by ecological systems.

    b. Act with restraint and efficiency when using energy, and rely increasingly on renewable energy sources such as solar and wind.

    c. Promote the development, adoption, and equitable transfer of environmentally sound technologies.

    d. Internalize the full environmental and social costs of goods and services in the selling price, and enable consumers to identify products that meet the highest social and environmental standards.

    e. Ensure universal access to health care that fosters reproductive health and responsible reproduction.

    f. Adopt lifestyles that emphasize the quality of life and material sufficiency in a finite world.

8. Advance the study of ecological sustainability and promote the open exchange and wide application of the knowledge acquired.

    a. Support international scientific and technical cooperation on sustainability, with special attention to the needs of developing nations.

    b. Recognize and preserve the traditional knowledge and spiritual wisdom in all cultures that contribute to environmental protection and human well-being.

    c. Ensure that information of vital importance to human health and environmental protection, including genetic information, remains available in the public domain.

III. SOCIAL AND ECONOMIC JUSTICE

9. Eradicate poverty as an ethical, social, and environmental imperative.

    a. Guarantee the right to potable water, clean air, food security, uncontaminated soil, shelter, and safe sanitation, allocating the national and international resources required.

    b. Empower every human being with the education and resources to secure a sustainable livelihood, and provide social security and safety nets for those who are unable to support themselves.

    c. Recognize the ignored, protect the vulnerable, serve those who suffer, and enable them to develop their capacities and to pursue their aspirations.

10. Ensure that economic activities and institutions at all levels promote human development in an equitable and sustainable manner.

    a. Promote the equitable distribution of wealth within nations and among nations.

    b. Enhance the intellectual, financial, technical, and social resources of developing nations, and relieve them of onerous international debt.

    c. Ensure that all trade supports sustainable resource use, environmental protection, and progressive labor standards.

    d. Require multinational corporations and international financial organizations to act transparently in the public good, and hold them accountable for the consequences of their activities.

11. Affirm gender equality and equity as prerequisites to sustainable development and ensure universal access to education, health care, and economic opportunity.

    a. Secure the human rights of women and girls and end all violence against them.

    b. Promote the active participation of women in all aspects of economic, political, civil, social, and cultural life as full and equal partners, decision makers, leaders, and beneficiaries.

c. Strengthen families and ensure the safety and loving nurture of all family members.

12. Uphold the right of all, without discrimination, to a natural and social environment supportive of human dignity, bodily health, and spiritual well-being, with special attention to the rights of indigenous peoples and minorities.

a. Eliminate discrimination in all its forms, such as that based on race, color, sex, sexual orientation, religion, language, and national, ethnic or social origin.

b. Affirm the right of indigenous peoples to their spirituality, knowledge, lands and resources and to their related practice of sustainable livelihoods.

c. Honor and support the young people of our communities, enabling them to fulfill their essential role in creating sustainable societies.

d. Protect and restore outstanding places of cultural and spiritual significance.

IV. DEMOCRACY, NONVIOLENCE, AND PEACE

13. Strengthen democratic institutions at all levels, and provide transparency and accountability in governance, inclusive participation in decision making, and access to justice.

a. Uphold the right of everyone to receive clear and timely information on environmental matters and all development plans and activities which are likely to affect them or in which they have an interest.

b. Support local, regional and global civil society, and promote the meaningful participation of all interested individuals and organizations in decision making.

c. Protect the rights to freedom of opinion, expression, peaceful assembly, association, and dissent.

d. Institute effective and efficient access to administrative and independent judicial procedures, including remedies and redress for environmental harm and the threat of such harm.

e. Eliminate corruption in all public and private institutions.

f. Strengthen local communities, enabling them to care for their environments, and assign environmental responsibilities to the levels of government where they can be carried out most effectively.

14. Integrate into formal education and lifelong learning the knowledge, values, and skills needed for a sustainable way of life.

a. Provide all, especially children and youth, with educational opportunities that empower them to contribute actively to sustainable development.

b. Promote the contribution of the arts and humanities as well as the sciences in sustainability education.

c. Enhance the role of the mass media in raising awareness of ecological and social challenges.

d. Recognize the importance of moral and spiritual education for sustainable living.

15. Treat all living beings with respect and consideration.

a. Prevent cruelty to animals kept in human societies and protect them from suffering.

b. Protect wild animals from methods of hunting, trapping, and fishing that cause extreme, prolonged, or avoidable suffering.

c. Avoid or eliminate to the full extent possible the taking or destruction of non-targeted species.

16. Promote a culture of tolerance, nonviolence, and peace.

a. Encourage and support mutual understanding, solidarity, and cooperation among all peoples and within and among nations.

b. Implement comprehensive strategies to prevent violent conflict and use collaborative problem solving to manage and resolve environmental conflicts and other disputes.

c. Demilitarize national security systems to the level of a non-provocative defense posture, and convert military resources to peaceful purposes, including ecological restoration.

d. Eliminate nuclear, biological, and toxic weapons and other weapons of mass destruction.

e. Ensure that the use of orbital and outer space supports environmental protection and peace.

f. Recognize that peace is the wholeness created by right relationships with oneself, other persons, other cultures, other life, Earth, and the larger whole of which all are a part.

THE WAY FORWARD

As never before in history, common destiny beckons us to seek a new beginning. Such renewal is the promise of these Earth Charter principles. To fulfill this promise, we must commit ourselves to adopt and promote the values and objectives of the Charter.

This requires a change of mind and heart. It requires a new sense of global interdependence and universal responsibility. We must imaginatively develop and apply the vision of a sustainable way of life locally, nationally, regionally, and globally. Our cultural diversity is a precious heritage and different cultures will find their own distinctive ways to realize the vision. We must deepen and expand the global dialogue that generated the Earth Charter, for we have much to learn from the ongoing collaborative search for truth and wisdom.

Life often involves tensions between important values. This can mean difficult choices. However, we must find ways to harmonize diversity with unity, the exercise of freedom with the common good, short-term objectives with long-term goals. Every individual, family, organization, and community has a vital role to play. The arts, sciences, religions, educational institutions, media, businesses, nongovernmental organizations, and governments are all called to offer creative leadership. The partnership of government, civil society, and business is essential for effective governance.

In order to build a sustainable global community, the nations of the world must renew their commitment to the United Nations, fulfill their obligations under existing international agreements, and support the implementation of Earth Charter principles with an international legally binding instrument on environment and development.

Let ours be a time remembered for the awakening of a new reverence for life, the firm resolve to achieve sustainability, the quickening of the struggle for justice and peace, and the joyful celebration of life.

BIBLIOGRAPHY

Ackerman, Frank, and Lisa Heinzerling. 2004. *Priceless: On Knowing the Price of Everything and the Value of Nothing.* New York: New Press.

Adger, W. Neil et al. eds. 2006. *Fairness in Adaptation to Climate Change.* Cambridge, MA: MIT Press.

Agarwal, Anil, and Sunita Narain. 1991. *Global Warming in an Unequal World: A Case of Environmental Colonialism.* New Delhi, India: Centre for Science and Environment.

Aitkinson, R. 2005. *Gentrification in a Global Context: The New Urban Colonialism.* Andover, UK: Routledge.

Allen, James et al. 2000. *Without Sanctuary: Lynching Photography in America.* Santa Fe, NM: Twin Palm Publishers, p. 15.

Allen, Patricia. 2004. *Together at the Table: Sustainability and Sustenance in the American Agrifood System.* University Park: Pennsylvania State University Press.

Allen-Gil, Susan et al. 2008. *Addressing Global Environmental Security through Innovative Educational Curricula.* New York: Springer.

Andersen, Steven O. et al. 2002. *Protecting the Ozone Layer: The United Nations History.* London, UK: Earthscan.

Anderson, William. 2001. *Economics, Equity, Environment.* Washington, DC: Environmental Law Institute.

Angotti, Tom. 2008. *New York for Sale: Community Planning Confronts Global Real Estate.* Cambridge, MA: MIT Press.

Atran, Scott, and Douglas Medin. 2008. *The Native Mind and the Cultural Construction of Nature.* Cambridge, MA: MIT Press.

Baber, Walter F., and Robert V. Bartlett. 2005. *Deliberative Environmental Politics: Democracy and Ecological Rationality.* Cambridge, MA: MIT Press.

Bacon, Christopher M. et al., eds. 2008. *Confronting the Coffee Crisis: Fair Trade, Sustainable Livelihoods and Ecosystems in Mexico and Central America.* Cambridge, MA: MIT Press.

Bartlett, Peggy, ed. 2005. *Urban Place: Reconnecting with the Natural World.* Cambridge, MA: MIT Press.

Beierle, Thomas C., and Jerry Cayford. 2002. *Democracy in Practice: Public Participation in Environmental Decisions.* Washington, DC: RFF Press.

Bell, Simon, and Stephen Morse. 2008. *Sustainability Indicators: Measuring the Immeasurable?* London, UK: Earthscan.

Benfield, F. Kaid et al. 2001. *Solving Sprawl: Models of Smart Growth in Communities across America.* Washington, DC: Island Press.

Benstein, Jeremy. 2006. *The Way into Judaism and the Environment.* Woodstock, VT: Jewish Light Publications.

Betsill, Michele M., and Felix Corell, eds. 2007. *NGO Diplomacy: The Influence of Nongovernmental Organizations in International Environmental Negotiations.* Cambridge, MA: MIT Press.

Blewitt, John. 2008. *Community Development, Empowerment and Sustainable Development.* Devon, UK: Green Books.

Bonorris, Steven, ed. 2007. *Supplemental Environmental Projects: A Fifty State Survey with Model Practices.* Chicago, IL: American Bar Association. PDF file available at no cost from the Web site of the Environmental Justice Committee of the Section of Individual Rights and Responsibilities, American Bar Association.

Bowler, I. et al. 2002. *The Sustainability of Rural Systems: Geographical Interpretations.* New York: Springer.

Braidotti, Rosi et al. 1994. *Women, the Environment and Sustainable Development: Towards a Theoretical Synthesis.* London; Atlantic Highlands, NJ: Zed Books in association with INSTRAW.

Brooks, Nancy, and Rajiv Sethi, Rajiv. "The Distribution of Pollution: Community Characteristics and Exposure to Air Toxics." *Journal of Environmental Economics and Management* 32 (1997): 233, 243–246.

Brophy, Alfred L. 2006. *Reparations: Pro and Con.* New York: Oxford University Press.

Buckingham, Susan, and Kate Theobald. 2003. *Local Environmental Sustainability.* Boca Raton, FL: CRC Publishing.

Bullard, Robert. 2000. *Dumping in Dixie: Race, Class, and Environmental Quality.* New York: Westview Press.

Bullard, Robert D., and Glenn S. Johnson, eds. 1997. *Just Transportation: Dismantling Race, and Class Barriers to Mobility.* Stony Creek CT: New Society Publishers.

Bullard, Robert D. et al. "Toxic Wastes and Race at Twenty: Why Race Still Matters after All of These Years." *Journal of Environmental Law* 38 (2008): 371.

Callicott, J. Baird. 1989. *In Defense of the Land Ethic: Essays in Environmental Philosophy.* New York: SUNY Press.

Cannavo, Peter F. 2007. *The Working Landscape: Founding, Preservation, And The Politics Of Place.* Cambridge, MA: MIT Press.

Capra, Fritjof. 1996. *The Web of Life: A New Scientific Understanding of Living Systems.* New York: Anchor Books.

Christensen, Julia. 2008. *Big Box Reuse.* Cambridge, MA: MIT Press.

Clapp, Jennifer, and Peter Dauvergne. 2005. *Paths to a Green World: The Political Economy of the Global Environment.* Cambridge, MA: MIT Press.

Cole, Luke. 2001. *From the Ground Up: Environmental Racism and the Rise of the Environmental Justice Movement.* New York: New York University Press.

Collier, Paul. 2009. *The Bottom Billion: Why the Poorest Countries Are Failing and What Can Be Done about It.* London, UK: Oxford Press.

Collin, Robert. 2006. *The Environmental Protection Agency: Cleaning Up America's Act.* Westport, CT: Greenwood Press.

Collin, Robert W. 2008. *Battleground: Environment.* Westport, CT: Greenwood Press.

Collin, Robert W. "Environmental Justice in Oregon: It's the Law." *Environmental Law* 38 (2008): 413.

Collin, Robin Morris, and Robert William Collin. "Where Did All the Blue Skies Go? Sustainability and Equity: The New Paradigm." *Journal of Environmental Law And Litigation* 9 (1994): 399–460.

Commoner, Barry. 1990. *Making Peace with the Planet.* New York: Pantheon Books.

Conca, Ken. 2005. *Governing Water: Contentious Transnational Politics and Global Institution Building.* Cambridge, MA: MIT Press.

Cutter, Susan L. 1993. *Living with Risk: The Geography of Technological Hazards.* London, UK; New York: E. Arnold.

Cutter, Susan L. 2006. *Hazards, Vulnerability and Environmental Justice.* London, UK; Sterling, VA: Earthscan.

De-Shalit, Avener. 1995. *Why Posterity Matters: Environmental Policies and Future Generation.* London, UK: Routledge.

De Sousa, Christopher. 2008. *Brownfields Redevelopment and Quest for Sustainability.* Oxford, UK: Elsevier Science.

Dietz, Thomas, and Paul C. Stern. 2008. *Public Participation in Environmental Assessment and Decision Making.* Washington, DC: National Academies Press.

DiMento, Joseph F. C., and Pamela Doughman, eds. 2007. *Climate Change: What It Means for Us, Our Children, and Our Grandchildren.* Cambridge, MA: MIT Press.

Dixon, Tom et al., eds. 2007. *Sustainable Brownfields Regeneration: Livable Places from Problem Spaces.* Hoboken, NJ: Wiley-Blackwell.

Devuyst, Dimitri et al. 2001. *How Green Is the City?: Sustainability Assessment and the Management of Urban Environments.* New York: Columbia University Press.

Dowies, Mark. 2009. *Conservation Refugees: The Hundred-Year Conflict between Global Conservation and Native Peoples.* Cambridge, MA: MIT Press.

Dubash, Novraz K., and Daniel Bouille. 2002. *Power Politics: Equity and Environment in Electricity Reform.* Washington, DC: World Resources Institute.

Edwards, Andres R. 2005. *The Sustainability Revolution: Portrait of a Paradigm Shift.* Gabriola Island, BC: New Society Publishers.

Elkins, Paul. 2000. *Economic Growth and Environmental Sustainability.* London, UK: Routledge.

Epler, Megan Wood. 2002. *Ecotourism: Principles, Practices, and Policies.* New York: United Nations Press.

Feagin, Joe R., and Karyn D. McKinney. 2003. *The Many Costs of Racism.* Lanham, MD: Rowman and Littlefield Publishers.

Field, John. 2006. *Social Capital*. Andover, UK: Routledge.

Folz, Richard C. et al., eds. 2003. *Islam and Ecology: A Bestowed Trust*. Cambridge, MA: Center for the Study of World Religions Harvard University Press.

Freyfogle, Eric T. 2003. *The Land We Share: Private Property and the Common Good*. Washington, DC: Island Press/Shearwater Books.

Freyfogle, Eric T. 1993. *Justice and the Earth: Images for Our Planetary Survival*. New York: The Free Press.

Fuchs, D. A. 2003. *An Institutional Basis for Environmental Stewardship: The Structure and Quality of Property Rights*. New York: Springer.

Gallagher, Kevin P., and Lyuba Zarsky. 2007. *The Enclave Economy: Foreign Investment and Sustainable Development in Mexico's Silicon Valley*. Cambridge, MA: MIT Press.

Geisler, Charles, and Gail Daneker, eds. 1997. *Property and Values: Alternatives to Public and Private Ownership*. Washington, DC: Island Press.

Gibson Robert B. et al. 2005. *Sustainability Assessment: Criteria and Process*. London, UK: Earthscan.

Gilbert, Charlene, and Eli Quinn. 2000. *Homecoming: The story of African American Farmers*. Boston, MA: Beacon Press.

Glazer, Nathan. 1998. *We Are All Multiculturalists Now*. Cambridge, MA: Harvard University Press.

Goldfield, David. 2006. *Encyclopedia of American Urban History*. Thousand Oaks, CA: Sage Publishing.

Gore, Al. 1992. *Earth in Balance: Ecology and the Human Spirit*. Boston, MA: Houghton Mifflin.

Gottlieb, Robert. 2004. *This Sacred Earth: Religion, Nature, Environment*. Andover, UK: Routledge.

Gottlieb, Robert. 2007. *Reinventing Los Angeles: Nature and Community in the Global City*. Cambridge, MA: MIT Press.

Gross, Julian. 2005. *Community Benefits Agreements: Making Development Projects Accountable*. San Francisco: Good Jobs First.

Grove, Richard. 1998. *Ecology, Climate and Empire: Colonialism and Global Environmental History, 1400–1940*. Isle of Harris, UK: White Horse Press.

Guber, Deborah Lynn. 2003. *The Grassroots of a Green Revolution: Polling America on the Environment*. Cambridge, MA: MIT Press.

Hamilton, Clive. 2004. *Growth Fetish*. London, UK: Pluto Press.

Haroff, Kevin T., and Katherin Kirwan Moore. "Global Climate Change and the National Environmental Policy Act. *University of San Francisco Law Review* 42, no. 1 (2007): 155.

Harris, Leslie M. 2002. *In the Shadow of Slavery: African Americans in New York City 1626–1863*. Chicago, IL: Chicago University Press.

Harris, William. 2009. *African American Community Development*. Unpublished manuscript, available from authors.

Hayword, Tim. 2005. *Constitutional Environmental Rights*. London, UK: Oxford University Press.

Hemmati, Minu et al. 2002. *Multi-Stakeholder Processes for Governance and Sustainability: Beyond Conflict and Deadlock*. London, UK: Earthscan.

Hess, David J. 2009. *Localist Movements in a Global Economy: Sustainability, Justice, and Urban Development in the US*. Cambridge, MA: MIT Press.

Hofrichter, Richard., ed. 2000. *Reclaiming the Environmental Debate: The Politics of Health in a Toxic Culture*. Cambridge, MA: MIT Press.

Jacobs, Jane. 1961. *The Death and Life of Great American Cities*. New York: Random House.

Jacobs, Jane. 2000. *The Nature of Economies*. New York: Vintage Books.

Jiggins, Janice. 1994. *Changing the Boundaries: Women-Centered Perspectives on Population and the Environment*. Washington, DC: Island Press.

Johnson, Steven M. 2004. *Economics, Equity and the Environment*. Washington, DC: Environmental Law Institute.

Johnston, Barbara Rose, and Susan Slyomovics, eds. 2008. *Waging War, Making Peace: Reparations and Human Rights*. Walnut Creek, CA: Left Coast Press.

Josephson, Paul. 2004. *Resources under Regimes: Technology, Environment, and the State*. Cambridge, MA: Harvard University Press.

Just, Richard E., and Sinaia Netanyahu. 1998. *Conflict and Cooperation on Trans-Boundary Water Resources*. New York: Springer.

Kaner, Sam et al. 2007. *Facilitator's Guide to Participatory Decisionmaking*. Hoboken, NJ: Jossey-Bass.

Karlof, Linda, and Terre Satterfield. 2005. *Earthscan Reader in Environmental Values*. London, UK: Earthscan.

Kasemir, Bernd, and Jill Jager. 2003. *Public Participation in Sustainability Science: A Handbook*. Cambridge, UK: Cambridge University Press.

Kazis, Richard. 1982. *Fear at Work: Job Blackmail, Labor, and the Environment*. New York: Pilgrim Press.

Kible, Paul Stanton. 2007. *Rivertown: Rethinking Urban Rivers.* Cambridge, MA: MIT Press.

Kusmer, Kenneth L, and Joe W. Trotter. 2009. *African American Urban History since World War Two.* Chicago, IL: Chicago University Press.

Lawn, Philipp Andrew. 2000. *Toward Sustainable Development: An Ecological Economics Approach.* Boca Raton, FL: CRC Press.

Lees, Loretta et al. *Gentrification.* Andover, UK: Routledge.

Maser, Chris. 1999. *Ecological Diversity in Sustainable Development: The Vital and Forgotten Dimension.* Boca Raton, FL: Lewis Publishers.

Medard, Gabel, and Henry Bruner. 2003. *Global Inc.: An Atlas of the Multinational Corporation.* New York: New Press.

Mission, C. F. 2005. *Sharing God's Planet: A Christian Vision for a Sustainable Future.* London, UK: Church House Publishing.

Montgomery, David R. 2007. *Dirt: The Erosion Of Civilizations.* San Francisco: University of California Press.

Morello-Frosch, Rachel, Manuel Pastor, and James Saad. "EJ and Southern California's Riskscape: The Distribution of Air Toxics Exposures and Health Risks among Diverse Communities." *Urban Affairs Review* 36 (2001): 551.

Moses, Marion. 1995. *Designer Poisons: How to Protect Your Health and Home from Toxic Pesticides.* San Francisco: The Pesticide Education Center.

Mowforth, Martin, and Ian Munt. 1998. *Tourism and Sustainability: New Tourism in the Third World.* Andover, UK: Routledge.

Murphy, Craig N. 2007. *The United Nations Development Programme: A History.* Cambridge, UK: Cambridge University Press.

Myers, Nancy J., and Carolyn Raffensperger, eds. 2005. *Precautionary Tools for Reshaping Environmental Policy.* Cambridge, MA: MIT Press.

Neal, Ruth, and April Allen. 1998. *Environmental Justice: An Annotated Bibliography.* Atlanta, GA: Environmental Justice Resource Center.

New Revised Standard Version Bible. 2008. *The Green Bible.* San Francisco: Harper Row.

Northcott, Michael S. 1996. *The Environment and Christian Ethics.* Cambridge, UK: Cambridge University Press.

Ohshita, Stephanie B. "The Scientific and International Context for Climate Change Initiatives." *University of San Francisco Law Review.* 42, no. 1 (2007): 1.

O'Neill, John, Alan Holland, Alan, and Andrew Light. 2008. *Environmental Values.* Andover, UK: Routledge

Orr, David. 1994. *Earth in Mind: On Education, Environment and the Human Prospect.* Washington, DC: Island Press.

Paarlberg, Robert. 2009. *Starved for Science: How Biotechnology Is Being Kept out of Africa.* Cambridge, MA: Center for the Study of World Religions, Harvard University Press.

Paavola, Jouni, and Ian Lowe. 2005. *Environmental Values in a Globalizing World: Nature, Justice, and Governance.* Andover, UK: Routledge.

Paehlke, Robert. 2008. *Democracy's Dilemma: Environment, Social Equity, and the Global Economy.* Cambridge, MA: MIT Press.

Pirages, Dennis, and Ken Cousins. 2005. *From Resource Scarcity to Ecological Security: Exploring New Limits to Growth.* Cambridge, MA: MIT Press.

Porter, Douglas. 2002. *Making Smart Growth Work.* Washington, DC: Urban Land Institute.

Putman, Robert D. 2004. *Democracies in Flux: The Evolution of Social Capital in Contemporary Society.* London, UK: Oxford University Press.

Rappaport, Roy A. 1994. "Human Environment and the Notion of Impact." In *Who Pays the Price? The Sociocultural Context of Environmental Crisis,* ed. Barbara Rose Johnston. Washington, DC: Island Press.

Raymond, Leigh Stafford. 2003. *Private Rights in Public Resources: Equity and Property Allocation in Market-Based Environmental Policy.* Washington, DC: Resources for the Future.

Rechtshaffen, Clifford, and Denise Anotoli. 2007. *Creative Common Law Strategies for Protecting the Environment.* Washington, DC: Island Press.

Revessz, Richard et al., eds. 2008. *Environmental Law, the Economy and Sustainable Development: The United States, the European Union, and the International Community.* Cambridge, MA: Cambridge University Press.

Riddell, Robert. 2004. *Sustainable Urban Planning: Tipping the Balance.* Malden, MA: Blackwell Publishing.

Robinson, Guy M. 2008. *Sustainable Rural Systems: Sustainable Agriculture and Rural Communities.* Aldershot, UK: Ashgate Publishing.

Robson, Mark G., and William E. Toscano. 2007. *Risk Assessment for Environmental Health.* Hoboken, NJ: Jossey-Bass.

Rogers, Heather. 2006. *Gone Tomorrow: The Hidden Life of Garbage.* New York: New Press.

Roodman, David Malin. 1996. *Paying the Piper: Subsidies, Politics, and the Environment.* Washington, DC: Worldwatch Institute.

Roseland, Mark. 2005. *Toward Sustainable Communities: Resources for Citizens and Their Governments.* Gabriola Island, BC: New Society Publishers.

Roszak, Theodore, Mary E. Gomes, and Allen D. Kanner, eds. 1995. *Ecopsychology: Restoring the Earth, Healing the Mind.* San Francisco: Sierra Club Books.

Sabatier, Paul et al., eds. 2005. *Swimming Upstream: Collaborative Approaches to Watershed Management.* Cambridge, MA: MIT Press.

Sachs, Jeffery. 2006. *The End of Poverty: Economic Possibilities for Our Time.* New York: Penguin Press.

Sagoff, Mark. 1988. *The Economy of the Earth: Philosophy, Law, and the Environment.* New York: Cambridge University Press.

Shiva, Vandana. 2005. *Earth Democracy: Justice, Sustainability, and Peace.* Boston, MA: South End Press.

Speth, James Gustave. 2008. *The Bridge at the Edge of the World: Capitalism, the Environment, and Crossing from Crisis to Sustainability.* New Haven, CT: Yale University Press.

Svedin, Uno Britt, and Hägerhäll Aniansson. 2002. *Sustainability, Local Democracy, and the Future: The Swedish Model.* Emeryville, CA: Kluwer Academic Publishers.

Sze, Julie. 2006. *Noxious New York: The Racial Politics of Urban Health and Environmental Justice.* Cambridge, MA: MIT Press.

Tilbury, Daniella, and David Wortman. 2004. *Engaging People in Sustainability.* Gland, Switzerland: International Union for Conservation of Nature and Natural Resources.

Tolley, Rodney. 2003. *Sustainable Transport: Planning for Walking and Cycling in Urban Environments.* Cambridge, UK: Woodhead Publishing.

United Nations. 2006. *Your Right to a Healthy Environment: A Simplified Guide to the Aaurhus Convention on Access to Information, Public Participation, in Decision Making.* New York: United Nations Publishing.

Weisman, Leslie Kanes. 2007. *Discrimination by Design: A Feminist Critique of the Man–Made Environment.* Champaign: University of Illinois Press.

Weiss, Thomas G. 2005. *UN Voices: The Struggle for Development and Social Justice.* Bloomington: Indiana University Press.

Whiteside, Kerry H. 2006. *Precautionary Politics: Principle and Practice in Confronting Environmental Risk.* Cambridge, MA: MIT Press.

Wildman, Stephanie M. 1996. *Privilege Revealed: How Invisible Preference Undermines America.* New York: New York University Press.

Wilkinson, Richard, and Kate Pickett. 2009. *The Spirit Level: Why More Equal Societies Almost Always Do Better.* St. Albans, UK: Allen Lane.

Williams, Juan. 2006. *Black Farmers in America.* Lexington: University of Kentucky Press.

Wilson, Edward O. 1984. *Biophilia.* Cambridge, MA: Harvard University Press.

World Bank. 2008. *The Global Monitoring Report.* World Bank, Washington, DC.

Zazueta, Aaron Eduardo. 1998. *Policy Hits the Ground: Participation and Equity in Environmental Policy Making.* Washington, DC: World Resources Institute.

Zovanayi, Gabor. *Growth Management for a Sustainable Future: Ecological Sustainability as the New Growth Management for the 21st Century.* Westport, CT: Greenwood Publishing.

INDEX

Boldface numbers refer to volume numbers. A key appears on all verso pages. Page numbers with a *f* following them indicate a figure.

ABOUT THE AUTHORS

ROBIN MORRIS COLLIN, professor of law, Willamette College of Law. Professor Morris Collin has taught at McGeorge School of Law, Tulane School of Law, Pepperdine Law School, Washington and Lee School of Law, University of Oregon School of Law, and Willamette School of Law. She has numerous publications in the area of sustainability and holds the David Brower Lifetime Achievement Award. Professor Collin was the first law professor to teach sustainability in the United States in 1993 and has taught it ever since. She has served as an advisor to state and federal environmental agencies. She has also litigated court cases and provided legislative testimony on many important environmental issues. She is currently working with the Oregon State Bar Association to find ways to integrate sustainability into legal practice. She is also appointed to the Oregon Environmental Justice Advisory Group.

ROBERT WILLIAM COLLIN is the senior research scholar at the Center for Sustainable Communities at Willamette University. He has been a professor of law, planning, and of social work, teaching at the University of Auckland, New Zealand; the University of Virginia Department of Urban and Environmental Studies; the University of Oregon's Department of Environmental Studies; Cleveland State University Department of Social Work; Jackson State University Department of Urban and Regional Planning; Lewis and Clark College of Law; and Willamette University College of Law. He has published many articles, book chapters, and book reviews. He has served as an advisor to state and federal environmental agencies and currently serves as chair of the Oregon Environmental Justice Advisory Group. His last two books are *The US Environmental Protection Agency: Cleaning up America's Act,* and *Battleground: Environment* (2 volumes).